Crime Lab Chemistry

Solving Mysteries with Chromatography

GEMS® Teacher's Guide for Grades 4–8

by
Jacqueline Barber and **Kevin Beals**
with **Carolyn Willard**

Skills
Observing • Inferring • Drawing Conclusions • Communicating
Designing an Investigation • Experimenting • Analyzing Data
Making Inferences • Evaluating Evidence

Concepts
Chemistry • Chromatography • Forensic Science • Separating Mixtures
Pigments • Dissolving • Solubility • Molecules • Capillary Action
Solvent • Medium • Test Substance

Themes
Systems and Interactions • Models

Mathematics Strands
Logic and Language • Discrete Mathematics

Nature of Science and Mathematics
Science and Technology • Real-Life Applications

Time
Two class sessions of 30–45 minutes;
three class sessions of 45–60 minutes

Great Explorations in Math and Science
Lawrence Hall of Science
University of California at Berkeley

Lawrence Hall of Science,
University of California,
Berkeley, CA 94720-5200

Initial support for the origination and publication of the GEMS series was provided by the A.W. Mellon Foundation and the Carnegie Corporation of New York. Under a grant from the National Science Foundation, GEMS Leaders Workshops were held across the United States. GEMS has also received support from: the Employees Community Fund of Boeing California and the Boeing Corporation; the people at Chevron USA; the Crail-Johnson Foundation; the Hewlett Packard Company; the William K. Holt Foundation; Join Hands, the Health and Safety Educational Alliance; the McConnell Foundation; the McDonnell-Douglas Foundation and the McDonnell-Douglas Employee's Community Fund; the Microscopy Society of America (MSA); the NASA Office of Space Science Sun-Earth Connection Education Forum; the Shell Oil Company Foundation; and the University of California Office of the President. GEMS also gratefully acknowledges the early contribution of word-processing equipment from Apple Computer, Inc. This support does not imply responsibility for statements or views expressed in publications of the GEMS program. For further information on GEMS leadership opportunities, or to receive a publications catalog and the *GEMS Network News,* please contact GEMS. We also welcome letters to the *GEMS Network News.*

International Standard Book Number: 0-924886-90-0

Printed on recycled paper with soy-based inks.

Library of Congress Cataloging-in-Publication Data

Barber, Jacqueline.
Crime lab chemistry : solving mysteries with chromatography / Jacqueline Barber, Kevin Rodney Beals, Carolyn Sandra Willard.
p. cm.
ISBN 0-924886-90-0 (trade paper)
1. Chromatographic analysis--Study and teaching (Elementary)--Activity programs. 2. Chemistry, Forensic. I. Beals, Kevin Rodney, 1959- II. Willard, Carolyn Sandra, 1947- III. Title.
QD79.C4B37 2004
363.25'62--dc22

2004020215

Acknowledgments

This new and expanded edition of *Crime Lab Chemistry* builds upon the original GEMS classic originated in the early 1980s by Jacqueline Barber, GEMS Director, when she was also Director of the LHS Chemistry Department. Jacquey was instrumental in helping shape and review this new edition. She and Cheryl Webb, GEMS Trial Test and Materials Manager, were also helpful in arranging a pilot-test site for the revised activities.

That excellent test site was Dana Wahlberg's 5th grade class at Oxford Elementary School in Berkeley, California. Dana provided suggestions and management tips, and her students inspired many modifications and improvements. Their inquiring faces appear in many of the guide's photographs.

Carolyn Willard, GEMS Network Director, served as GEMS co-author for this new edition, providing invaluable input during the pilot-testing process, and helping ensure the classroom applicability and reasonable length of the new unit.

The authors also wish to thank:

- UC Berkeley Chemistry Professor Angelica Stacy for her ideas and comments on the large-scale chromatography modeling demonstration involving a fan in Activity 2.
- Physicist Bruce Birkett for his scientific review of the unit.
- LHS educator Lynn Barakos for her help with the molecular models that students critique.
- LHS educator Catherine Halversen for suggestions about class discourse, reflection, and discussion during Activity 3.
- Jan Coonrod, of the LHS high school chemistry program Living by Chemistry, for her help with the molecular models and for her many salient comments concerning scientific accuracy. ■

Contents

Time Frame

Based on classroom testing, the following are guidelines to help give you a sense of how long the activities may take. The sessions may take less or more time with your class, depending on students' prior knowledge, their skills and abilities, the length of your class periods, your teaching style, and other factors. Try to build flexibility into your schedule so that you can extend the number of class sessions if necessary.

Activity 1: Laboratory Investigations

Session 1: Investigating the Evidence .... 45–60 minutes
Session 2: Solving the Mystery .... 30–45 minutes

Activity 2: What is Chromatography? .... 30–45 minutes

Activity 3: Designing Your Own Tests

Session 1: Choosing a New Test Substance .... 45–60 minutes
Session 2: Designing Chromatography Systems .... 45–60 minutes

What You Need for the Whole Unit

The quantities below are based on a class size of 32 students. You may, of course, require different amounts for smaller or larger classes. This list gives you a concise "shopping list" for the entire unit. Please refer to the "What You Need" and "Getting Ready" sections for each individual activity, which contain more specific information about the materials needed for the class and for each team of students.

Options for gathering your classroom materials for this unit:

- Some teachers prefer to gather materials needed to teach a GEMS unit themselves at local stores.

- Others prefer to purchase a ready-made GEMS Kit®.

The items marked with an asterisk in the list below are included in the commercial GEMS Kit for this unit as of the time of publication. Check with the kit supplier to confirm what materials are included. Please see the inside front cover for more information on purchasing GEMS Kits.
(* provided in the GEMS Kit)

Nonconsumables

- * ❑ 9 or more different black felt-tip pens (see "Getting Ready" #5 on page 10)
- * ❑ 1 copy of the **Teacher Script** (pages 21–22 or 24–25, depending on which mystery scenario you choose; see "Getting Ready" #1 on page 9)
- * ❑ 1 copy of the **Suspect Statements** (page 23 or 26, depending on which mystery scenario you choose)
- * ❑ 6–8 water troughs (plastic wallpaper-paste troughs)
- ❑ about 5 small rocks, any kind, from about 1"–3"
- ❑ about 5 styrofoam balls, about 2" in diameter
- ❑ about 5 styrofoam packing "peanuts" (or other lightweight items)
- ❑ a large fan or blow dryer
- * ❑ about 25 wide-mouthed cups
- ❑ several pennies for crushing plant pigments onto paper strips
- ❑ 1 overhead transparency of each of the following:
    - * __ **Different Water Molecule Models** (page 27)
    - * __ **Model A** (page 38)
    - * __ **Model B** (page 39)
    - * __ **Model C** (page 40)
    - * __ **Model D** (page 41)
    - * __ the **Four Models** student sheet (pages 42–43)
- * ❑ 1 copy each of the **Candy Coating** (page 58), **Food Coloring** (page 59), and **Plants** (page 60) **Test Substance Procedure** sheets

* ❑ *(optional)* 8–16 magnifying lenses
❑ *(optional)* an extension cord for the fan or blow dryer

Consumables

❑ water
❑ 100 *white,* institutional paper towels⋆ (or large round white coffee filters or white household paper towels)—for Activity 1, Session 1
* ❑ about 200 4" strips of chromatography paper⋆ (or white coffee filters cut into 1" x 4" strips)—for Activity 3
* ❑ white coffee filters
❑ different brands of white paper towel
* ❑ white cotton cloth
* ❑ white construction paper
❑ toilet paper or tissues
❑ white notebook paper
❑ at least 12 brown and green M&M's®, Reese's Pieces®, or other candies with colored coatings
* ❑ a 1 oz. squeeze⋆⋆ bottle of green food coloring
* ❑ a 1 oz. squeeze⋆⋆ bottle of red food coloring (for mixing with green to make brown)
❑ a few different vegetable leaves such as red cabbage, spinach, beet greens, or other dark green leafy vegetables
❑ a few freshly picked deep green or red leaves from trees
❑ a few freshly picked flower petals that are a deep reddish brown
❑ newspaper or plastic trash bags to cover the surface of the Test Substances station
* ❑ 1 tablespoon of baking soda
* ❑ 1 tablespoon of salt
* ❑ 1 quart or liter bottle of white vinegar
* ❑ 1 quart or liter bottle of rubbing alcohol
❑ 32 copies of the **Four Models** student sheet (pages 42–43)
❑ 32 copies of the **New Microscope Eyes Model of Chromatography** student sheet (page 44)
❑ 16 copies of the **Our Chromatography Test** sheet (page 61)
❑ 32 copies of the **Chromatography Test Report** sheet (page 62)
❑ *(optional)* chromatography paper⋆—for Activity 1, Session 1
❑ *(optional)* other colors of M&M's or Reese's Pieces
❑ *(optional)* additional colors of food coloring
❑ *(optional)* clear plastic wrap to cover the rubbing alcohol cups

⋆In Activity 1, the use of chromatography paper is optional. Chromatography paper is specifically designed for paper chromatography. It presents the colors more vividly than paper towels or coffee filters, but is not necessary for Activity 1. In fact, many teachers think white paper towels or coffee filters are preferable for use with the pens in the first activity, because the process happens faster than with chromatography paper and the colors are clear enough.

However, for Activity 3 we recommend using chromatography paper, if possible, because it provides a much better separation, is more effective, and easier to interpret. Some of the subtle colors of candy coatings and plant pigments are difficult to notice without using it. Chromatography paper can be purchased from many scientific supply stores, including Carolina Biological Supply, the distributor of GEMS Kits. For more on obtaining chromatography paper, see page 85.

**If squeeze bottles of food coloring are not available, you can pour a very small amount into a cup and add a medicine dropper or a toothpick for students to use to apply the food coloring to the paper strip.

General Supplies

- * ❑ 2 or more different brands of brown water-based felt-tip pens
- * ❑ a variety of other colors of water-based felt-tip pens, especially green or purple
- ❑ 6 sharpened pencils for labeling paper towel strips at pen stations
- * ❑ 32 pencils (or drinking straws) long enough to set across the water trough
- ❑ 32 pencils
- ❑ several sheets of white scratch paper (onto which chromatograms will be taped)
- ❑ 32 pieces of unlined paper for students to draw on (in Activity 1, Session 2)
- ❑ about 5 wadded up individual sheets of 8 ½" x 11" paper
- ❑ a few Post-it® Notes or scratch paper to label the vegetables and other plant samples at the Test Substances station
- ❑ 1 ruler
- ❑ several pairs of scissors
- * ❑ masking tape
- ❑ an overhead projector
- ❑ *(optional)* 32 blank sheets of 8 ½" x 11" paper (for the "Questions" sheet used at end of Activity 2)
- ❑ *(optional)* a few different colored permanent pens

INTRODUCTION

This New GEMS edition of *Crime Lab Chemistry: Solving Mysteries with Chromatography* takes a classic GEMS activity and deepens its educational value by strengthening the science content and student inquiry abilities. New in this edition are multiple experiences for students to visualize the molecular nature and behavior of matter, as they create and revise their own models and consider the advantages and limitations of models. The new unit maintains and enhances exceptional student motivation and adds two updated mysteries to the mix. It presents all this in a manageable and time-efficient unit, somewhat longer than the original two-session guide, but still concise and much more content rich.

Crime Lab Chemistry draws upon students' interest in and enthusiasm for solving mysteries to convey important scientific concepts, methods, and techniques. Chromatography is one of the most important techniques in analytical chemistry. It is used for separating mixtures and has applications in many other scientific disciplines. In crime labs, chromatography is used to separate the components of "clue" substances such as blood, ink, gases, or other mixtures found at the scene of a crime. Chromatography can also provide an introductory window into the molecular theory of matter and, in this unit, this approach empowers students by giving them a chance to originate, discuss, and refine their own molecular models and scientific explanations of evidence.

In all the various types of chromatography, a test substance is placed onto or into a medium, and then a solvent is passed through or over it. As the solvent passes through the test substance, some of the test substance may be attracted to the solvent and follow it up the medium. Different types of molecules are transported different distances, causing them to separate.

A summary of different types of chromatography is included in the "Background for the Teacher" section on page 65.

The test substance in this unit's opening activity is ink. Ink is a mixture. Inks are made by mixing different ingredients, including pigments (chemicals that make something appear a particular color). Black inks are made by mixing a wide variety of pigments, and each type of black ink is usually made up of different and distinctive pigments.

In the first activity of this unit, students use chromatography to separate out the different pigments in the ink of black felt-tip pens. Paper towel is the medium, water is the solvent, and black ink is the test substance. When one edge of an absorbent paper is dipped in water, the water begins to creep up the paper through capillary action. As the water passes through a mark made by a black non-permanent pen on the paper, it will separate out the various pigments that make up the black ink. Because the molecules of each pigment vary in shape, size, weight, and in how much they are attracted or not attracted to the solvent or the medium (the water or the paper) each pigment is carried by the water to a different height on the paper. This makes a particular pattern of colors. The end result is that the inks your students test separate into bands of brightly colored pigments.

Evidence is a key idea in this unit and is central to all of science. It of course also occupies center stage in forensic science, criminal investigation, and the courtroom! The ability to distinguish between evidence and inference is not only essential to developing scientific literacy, it has enormous real-life applications and benefits. Reading a newspaper article or website promotion, evaluating a political candidate's claims and record, considering the environmental impact of a new building's construction, serving on a civil or criminal jury—these are only a few of the many instances where we depend on the ability to define evidence, analyze, evaluate, and weigh it, make inferences based on that evidence, and/or determine whether or not there is insufficient evidence to make a reasonable inference or conclusion.

Crime Lab Chemistry and National Content Standards

This unit focuses on one of the most important techniques in chemistry, chromatography, as well as on initial understandings of the particulate theory of matter. This places it firmly within the Physical Sciences content area of the *National Science Education Standards*. The unit also provides an excellent platform for distinguishing evidence from inference, creating and refining models, engaging in other aspects of scientific inquiry, and gaining understandings about technology. One major element of the "Unifying Concepts and Processes" component of the *National Standards* is summarized as "evidence, models, and explanation" and these underpin this unit. The following standards-related concepts and abilities are addressed in the New GEMS *Crime Lab Chemistry*:

Physical Science/Properties

- Grades K–4: Properties of objects and materials. Objects can be described by the properties of the materials from which they are made, and those properties can be used to separate or sort a group of objects or materials.

- Grades 5–8: Properties and changes of properties in matter. A mixture of substances often can be separated into the original substances using one or more of the characteristic properties.

- Grades 9–12: Structure and properties of matter. The understanding of the microstructure of matter can be supported by laboratory experiences with the macroscopic and microscopic world.

Note: In the *National Standards* the microstructure of matter is placed in Grades 9–12 while *Benchmarks for Science Literacy* calls for basic understanding of atomic structure by the end of 8th grade. In any case, researchers agree that prior concrete experiences (such as the activities in this unit) linked with models, evidence, and explanations, can help overcome early misconceptions and assist students in establishing a foundation for more complex understanding in later grades.

Scientific Inquiry/Abilities Necessary to Do Scientific Inquiry

- Use appropriate tools and techniques to gather, analyze, and interpret data.

- Develop descriptions, explanations, predictions, and models using evidence.

- Think critically and logically to make the relationships between evidence and explanations.

- Recognize and analyze alternative explanations and predictions. Students should develop the ability to listen to and respect the explanations proposed by other students. They should remain open to and acknowledge different ideas and explanations, be able to accept the skepticism of others, and consider alternative explanations.

Science and Technology

- Science and technology are reciprocal. Science helps drive technology, as it addresses questions that demand more sophisticated instruments and provides principles for better instrumentation and technique. Technology is essential to science, because it provides instruments and techniques that enable observations of objects and phenomena that are otherwise unobservable due to factors such as quantity, distance, location, size, and speed. Technology also provides tools for investigations, inquiry, and analysis.

Activity-by-Activity Overview

Activity 1: Laboratory Investigations consists of two class sessions. Before beginning the first session, you'll need to determine which of the two included mystery scenarios you'd like to use, if you'd like to make up your own, or if you'll present the activity without the mystery context. In **Session 1: Investigating the Evidence,** your students assume the role of crime lab chemists attempting to determine which black pen was used to write a ransom note. Students use the technique of paper chromatography to test several black felt-tip pens. In **Session 2: Solving the Mystery,** students compare the results of the pen tests with the results of the "ransom note" test and attempt to identify the "mystery pen" used to write the note. They discuss whether this evidence is enough to convict the suspect whose pen matched the note. After students learn about molecules, they imagine they have "microscope eyes" and draw a model of what they think may be happening to ink molecules in a chromatogram.

In **Activity 2: What is Chromatography?** students learn that chromatography is a system for separating mixtures, and the system consists of a medium, a solvent, and a test substance. The teacher demonstrates a model chromatography system, in which the test substance is represented by rocks, paper wads, styrofoam balls, and styrofoam "peanuts." The medium is represented by the floor, and the solvent by wind from a fan or hair dryer. Next students analyze and discuss four drawings that are possible models for how ink molecules separated in the chromatography test. Finally, given this new information, students evaluate and adjust their own molecular models.

Like the first activity, **Activity 3: Designing Your Own Tests** has two class sessions. In **Session 1: Choosing a New Test Substance,** students test a variety of test substances, such as colored pens, food coloring, leaves, flower petals, and food colorings on candies. In **Session 2: Designing Chromatography Systems,** students choose one test substance that interests them, then try out different combinations of mediums and solvents with their test substance until they find what they think is the best system for separating the pigments. Students write a report in which they explain their results and draw a new molecular "microscope eyes" model that shows what they think happened to the molecules of the test substance, medium, and solvent in their best chromatography system.

The Structure of Matter

The structure of matter has always been a subject of fascination. Early philosophers and proto-scientists from many cultures visualized some interesting ideas about atoms and molecules. As noted in the summary of how this unit aligns with the *National Standards,* a more complete understanding of the particulate theory of matter, atomic structure, and related concepts is most appropriate at higher grades, from high school onward into college, graduate school, and high-level research. However, students hear about atoms and molecules at very early ages, and begin developing their own ideas about what these words and concepts might mean. Age-appropriate instruction can be extremely helpful in preparing the ground for later grasp of more complex ideas, and this unit has been revised with that goal in mind. See "Background for the Teacher" for a brief description of science education research findings on this subject. One summary concludes, "there is some evidence that carefully designed instruction carried out over a long period of time may help middle-school students develop correct ideas about particles."

Kyle told me Tuesday in the car on the way home that "it" (Crime Lab Chemistry) was the best thing they'd done so far this year. I asked why and he said, "because we didn't have to learn anything!" I asked what the kids were doing all that time if they weren't learning. He said, "trying to figure out how molecules of water move," and then gave an elaborate explanation. When I said, "hmm...sounds like you were learning to me," he said, "oh, you know, not learning all that junk we usually do." He then described at great length the crime, the mystery, the pens, the letters, the ink, the colors, etc. It was very different than the usual "aw-nothing-much" answers I often get about what they've done at school.

—A parent of a student in a classroom where this New GEMS guide was tested

The goal of the unit is to open up students' thinking about the particulate theory of matter, and provide them with an interesting series of chemistry-related activities and challenges to help anchor their conceptual thinking. Students start thinking about the structure of matter and realizing that models of matter's structure are one way of understanding interesting phenomena in nature and science. The goal is *not* to bring all students to full understanding or manage to overcome in one short unit the misunderstandings that even many adults have about atoms and molecules.

The GEMS *Dry Ice Investigations* unit, for Grades 6–8, also combines intriguing physical phenomena with molecular models and an age-appropriate introduction to the particulate theory of matter. In addition, it carefully scaffolds a wide range of guided and full inquiry opportunities. It would strongly complement this *Crime Lab Chemistry* unit.

The GEMS unit Investigating Artifacts *also makes a strong complement to these mystery units, as it explores issues of evidence and inference within the context of learning about Native American and world cultures and archaeology.*

And, if your students like being crime lab scientists, after *Crime Lab Chemistry* consider presenting the GEMS unit *Fingerprinting,* then crown the experience with *Mystery Festival.* These units are motivating to students, foster important skills and techniques, and help students evaluate evidence and inference. They also convey a larger lesson: that the process of detecting, collecting clues, and seeking to solve a mystery is a big part of what science is all about! ■

Overview

Activity 1 consists of two class sessions. In the first session, students are challenged to solve a mystery by using a scientific test to determine who wrote a "ransom note" with a black felt-tip pen. They begin by testing a blank paper strip in water, and observe the water moving up the paper. They then use the same technique on a piece of the ransom note and observe as the colorful pigments are separated as the water moves up the paper. They learn that chromatography is a technique that separates substances from a mixture. They then use this technique to test the inks from the suspects' pens to try to find out if there is a match to the pen used for the ransom note. Later, in the second session, students will analyze and discuss their results and begin drawing a model of what is happening at the molecular level.

In the second session, student pairs study the chromatograms from the pens they tested in the first session, and decide if any match the ransom note. They present their evidence to each other, and after a short discussion the class decides which pen was used for the ransom note. They debate and vote on whether or not the evidence from their chromatography tests alone is enough to convict the suspect with the matching pen.

After another careful examination of their chromatograms, students focus on the different pigments in the inks and how they traveled up the paper. After learning about molecules, students are challenged to pretend they have "microscope eyes" and to imagine they can see what water, ink, and paper molecules look like. They draw a model to show why some ink molecules go up the paper better than others. They evaluate the strengths and weaknesses of their models, and record this for future reference.

"Microscope eyes" suggests that students visualize things that are too small to see with the naked eye, as a microscope does. However, for the purposes of scientific accuracy, you may want to point out that even the most powerful light microscopes do **not** *allow people to see atoms because atoms are so small. Scanning tunneling electron microscopes can detect atoms, but that results in a computerized image, not a direct visualization. If you wanted to be more accurate, and introduce your students to a rapidly emerging field of science, you could discuss nanoscience (a nanometer is one-billionth of a meter). Then if students are asked to visualize things so small not even a microscope can see them, they could use their "nanovision"!*

Learning Objectives for Activity 1

Session 1

- Introduce students to chromatography as a technique in chemistry through which substances in a mixture can be separated.
- Set the stage for learning fundamental chemistry and forensic concepts and processes in the rest of the *Crime Lab Chemistry* unit.

Session 2

- Develop students' abilities to observe and discuss test results.
- Help students make inferences and draw conclusions based on evidence (and refrain from jumping to conclusions!).
- Enable students to understand that all substances are made of molecules, and their molecular structure in part determines how they interact with other substances.
- Have students begin to formulate a scientific model to explain chromatography.
- Help students understand that models can be very useful but also have limitations and that all models are inaccurate in one or more ways.

Session 1: Investigating the Evidence

■ What You Need

For the class:

- ❑ 1 copy of the **Teacher Script** (pages 21–22 or 24–25, depending on which mystery scenario you choose; see "Getting Ready" #1)
- ❑ 1 copy of the **Suspect Statements** (page 23 or 26, depending on which mystery scenario you choose)
- ❑ 100 *white,* institutional paper towels (or large round white coffee filters or white household paper towels)
- ❑ 9 or more different black felt-tip pens (see "Getting Ready" #5)
- ❑ 6 sharpened pencils for labeling paper towel strips
- ❑ water
- ❑ 1 ruler
- ❑ 1 pair of scissors
- ❑ *(optional)* chromatography paper*

*This is paper specifically designed for paper chromatography. It presents the colors more vividly than paper towels or coffee filters, but is not necessary for this activity. In fact, many teachers think white paper towels or coffee filters are preferable for use with the pens in this first activity, because the process happens faster than with chromatography paper and the colors are clear enough. Chromatography paper is most useful in Activity 3 when students investigate mixtures with pigments that are more difficult to see. See page 85 for information about purchasing chromatography paper.

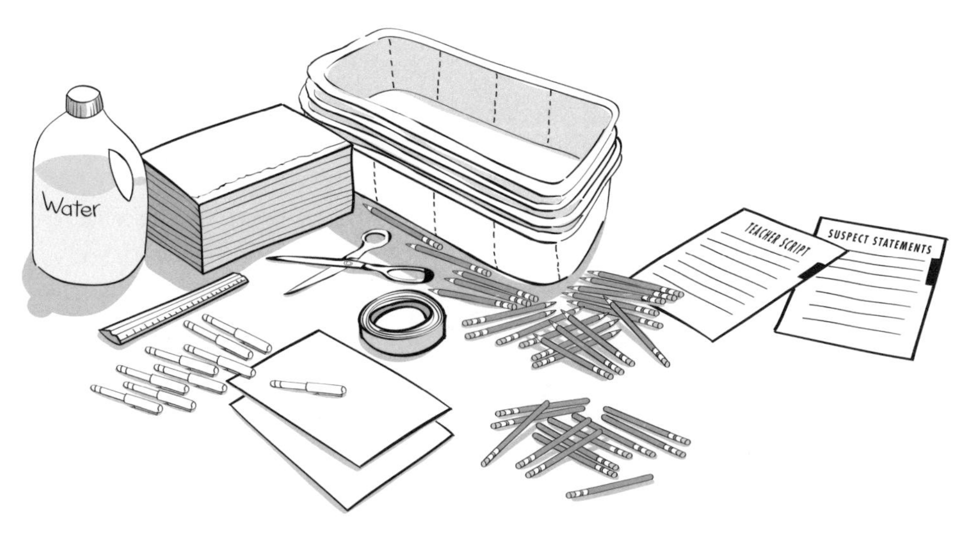

For each group of four students:

- ❑ 1 water trough (plastic wallpaper-paste trough)
- ❑ 2 sheets of white scratch paper
- ❑ 4 pencils (or drinking straws) long enough to set across the water trough
- ❑ about 2 feet of masking tape
- ❑ *(optional)* 1 or 2 magnifying lenses

■ Getting Ready

1. **Choose a mystery scenario.** Introducing the *Crime Lab Chemistry* unit with a make-believe mystery is fun and engaging for students. We have included two whimsical mystery scenarios for you to choose from on pages 21–22 and 24–25: **The Mystery of the Missing Cold Remedy** and **The Mystery of the Missing Dinosaur.** The session instructions indicate when to introduce the mystery and how to introduce the suspects. Although this teacher's guide is written with these scenarios in mind, you could also choose one of the following options:

    - Adapt one of the scenarios to your situation by changing suspect names and/or other details. Some teachers create in-school mysteries in which school staff (if they are willing to do so) are the suspects.

    - Make up your own mystery entirely.

    - Conduct the activity without using a mystery or suspects at all. (This option is not as motivating.)

2. Make one copy each of the **Teacher Script** (pages 21–22 or 24–25) and **Suspect Statements** sheet (page 23 or 26) for the mystery scenario you have chosen. Cut apart the six statements for students to read aloud.

3. **Prepare the chromatography materials.** Begin with the **paper strips.** Cut the paper towels (or coffee filters) into strips about 1" x 4". Depending on the depth of your water troughs, you may need to make strips a different length. **Test one before you cut them all.** When the strip is suspended from a pencil, the lower end of the strip (about $^1/_2$ inch) should hang in the water. You'll need roughly 125 to 150 strips for Activity 1.

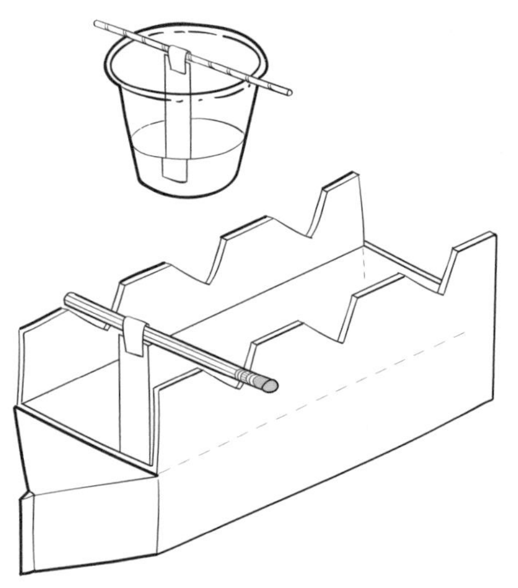

4. Next obtain and prepare the **water troughs.** The wallpaper troughs (about 15" x 6" x 3" or about 30" x 6" x 6") are inexpensive and available at paint or wallpaper supply stores. Cups, aluminum trays, or cut milk cartons (simply cut lengthwise, or as shown in the illustration) can also be used. All you need is a way to suspend strips of paper so the ends of the strips hang in water. If using cups, each group of four students would need about six wide-mouthed cups (two paper strips can hang into each cup). Pour no more than $^1/_2$ inch of water into the bottom of the wallpaper troughs. If necessary, move tables so four students can be seated around a water trough.

5. Gather and test **black pens.** Choose a variety of brands, including a pen with permanent ink. A collection of nine different black pens should provide at least six distinctly different patterns. To select the six pens to use in the mystery, you will need to conduct a chromatography test with *each* of the nine pens, using the following procedure:

Be sure to include the permanent pen as one of the six black pens to use in the mystery. Because its ink isn't soluble in water, it makes an interesting contrast to the other pens and leads to an introduction of solvents in Activity 2.

a. Draw a horizontal line across a paper strip about one inch from the bottom.

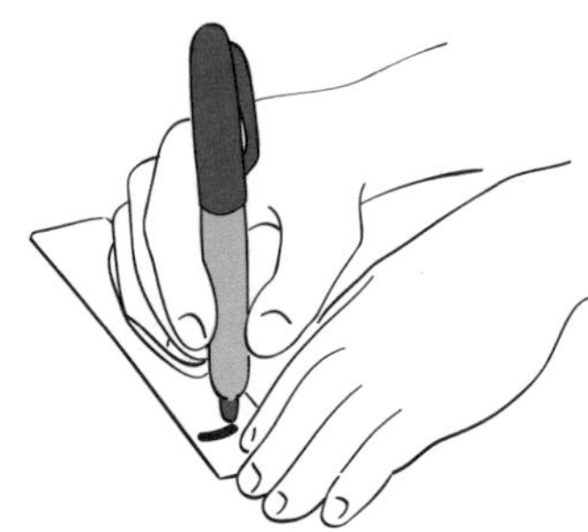

b. Use a pencil (NOT a pen) to write an abbreviated name for the pen at the top of the paper strip so you don't forget which strip goes with which pen.

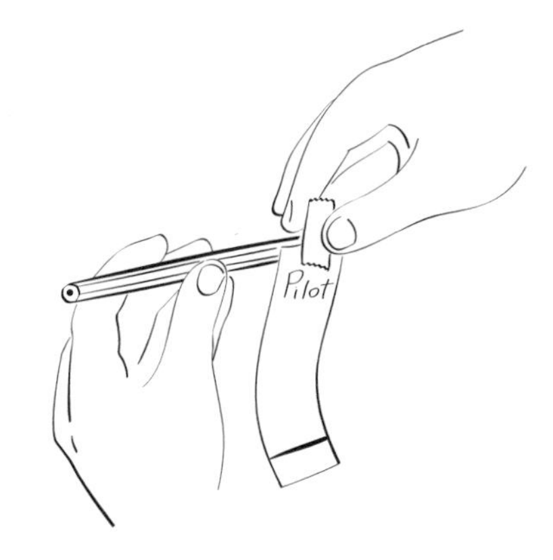

c. Attach the top of each strip to a pencil with tape. You may need to use two pencils so that all the strips can hang in the water at once.

d. Set the pencils on the trough so the paper strips hang down and the ends touch the water. **Make sure the ink mark remains at least $^1/_2$ inch *above* the water.**

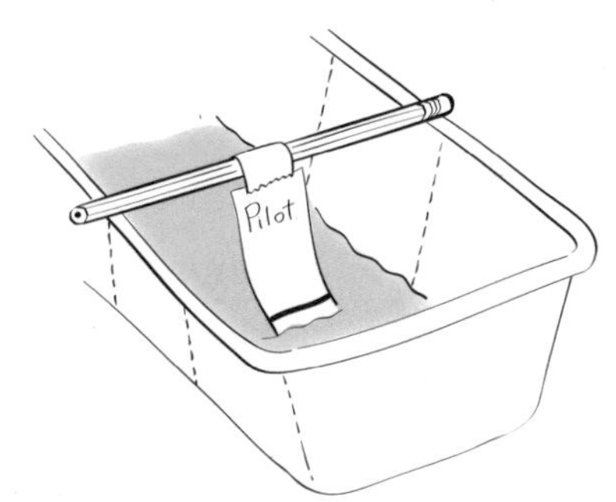

e. Remove each paper when the water has traveled about **three-quarters of the way up the strip.**

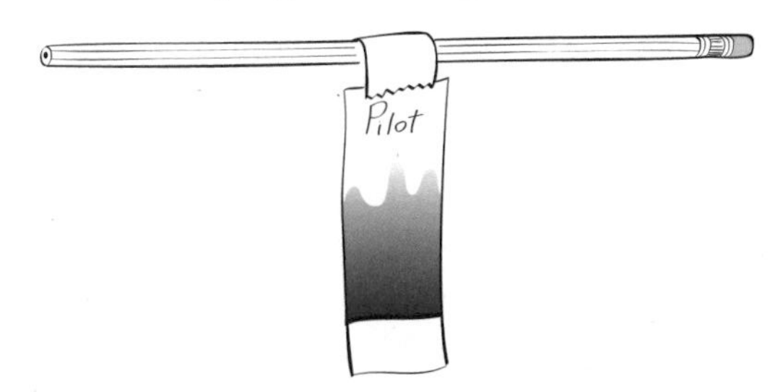

6. **Select and number the pens.** After the paper is dry, compare the test strips of all nine pens and choose the six you'll use as the suspect's pens. Remember you're looking for distinctly different patterns. Purchase two of each brand of pen (for a total of 12 pens). Of these, select one brand with a unique color pattern to be the "mystery pen." Decide which of the six suspects will be "guilty," and assign the mystery pen to your "guilty" suspect. (For older students, choose a mystery pen that has a color pattern only subtly different from that of some of the other pens. For younger students, make sure that the difference is more obvious.)

   a. From the **Suspect Statements** sheet you cut into strips, get the number of the suspect you have chosen to be guilty. Put that number on the barrel of the mystery pen. Then put the numbers of the other suspects on the barrels of the remaining five black pens.

   b. On the board (or butcher paper or overhead transparency) write a list of the six suspects' names and numbers. This list should be visible to the entire class and will need to be referred to periodically during Activities 1 and 2.

c. At six different places around the room, set up one station for each brand of pen. Place two pens at each station. This set up will allow many students to use the pens at one time. Also place a sharpened pencil at each station. Students will use it to label their paper strip with the number of the pen.

7. Prepare pieces/strips of the **ransom note.** There are two ways to do this.

   a. One option is to start with a ransom note in one piece, writing it with the mystery pen. To do this, cut a white paper towel to a height of four inches and write a line of text from the ransom note (see the **Teacher Script**) on it, across the paper towel, about one inch up from the bottom. Keep this handy during class so just after you introduce the activity you can cut the note into strips while the students watch. You'll have to cut enough for each pair of students to have one strip (plus a few extra).

   OR

   b. Another option is to prepare some paper strips beforehand. Use the mystery pen to draw a black line about one inch up from the bottom of some paper strips. You can then tell students during class that the strips are "pieces of the original ransom note" on which they will conduct a chromatography test. Make enough for each pair of students to have one strip (plus a few extra).

8. **Put it all together.** Have the following ready for quick distribution after you introduce the activity: the ransom note you prepared (or the strips with black lines on them), blank paper strips, sheets of scratch paper, pencils, tape, and water troughs. Also have the numbered list of suspects ready to unveil after you've introduced them.

9. If you've decided to use them, set out the magnifying lenses.

■ Introducing the Mystery

1. From the **Teacher Script** you've selected, read aloud "Introducing the Mystery," including the ransom note.

If you have chosen not to use a mystery, simply tell the class that the police need assistance in determining who might have written a ransom note.

2. Tell your students the ransom note was written on white paper towel with a black felt-tip pen. Six suspects with black felt-tip pens have been identified and their pens seized. Point out the pens you set out earlier. *Do not introduce the suspects yet.*

3. Ask for some ideas of how students could figure out which of the pens, if any, was used to write the ransom note. Accept all ideas. [Do a handwriting analysis, take fingerprints, analyze the ink, look for DNA evidence, etc.]

4. Write the word ***chromatography*** on the board and explain that it's a technique chemists use in crime labs. Say you will show them how to use chromatography to help solve the mystery. Tell them that later you will give them pieces of the ransom note to use in their investigations, but first they'll conduct a blank test.

■ Conducting a Blank Test

1. Begin by using a *blank* paper strip to demonstrate the technique.

   a. Attach the blank paper strip to a pencil with tape.

   b. Set the pencil on the trough, so the paper strip hangs down just touching the water.

   c. Point out that they should be careful to avoid knocking the pencil into the trough.

2. Have students predict what will happen. Encourage brief discussion, giving students a chance to explain their reasoning.

3. Divide students into groups of four. Distribute a water trough, two pencils, and some tape to each group and let students know they'll do what you just demonstrated. Pass out a blank test strip to each *pair* of students, and have them begin. Circulate and ask them to describe exactly what they observe happening. [They will observe water traveling up the paper.]

This blank test may lack the excitement of the ones that follow, but it is important to help focus their attention on the fact that the water is moving up the paper. As groups begin to see the results, you might want to ask them to try to explain why the water travels up the paper. This can be an interesting discussion topic. (See "Background for the Teacher" for more information on capillary action.)

■ Testing the Ransom Note

1. Read the ransom note aloud again.

2. If you've written a line of the ransom note on a paper towel, cut off a strip and hold it up for the students to see. Alternatively, if you've already prepared paper strips, hold one of them up.

3. Invite the students to predict what will happen as water travels up the paper through the ink marks. Record their predictions on the board. You may want to mention that scientists often make predic-

tions about what they think might happen. Encourage active class participation.

4. Remind the students that this fragment of ransom note is the *only* piece of evidence they have to work with. Emphasize the following cautions to be taken when conducting a chromatography test on the evidence.

   a. Do not let the ink line dip below the surface of the water.

   b. Remove the paper when the water has traveled about three-quarters of the way up the strip, and tape it to a sheet of scratch paper to dry.

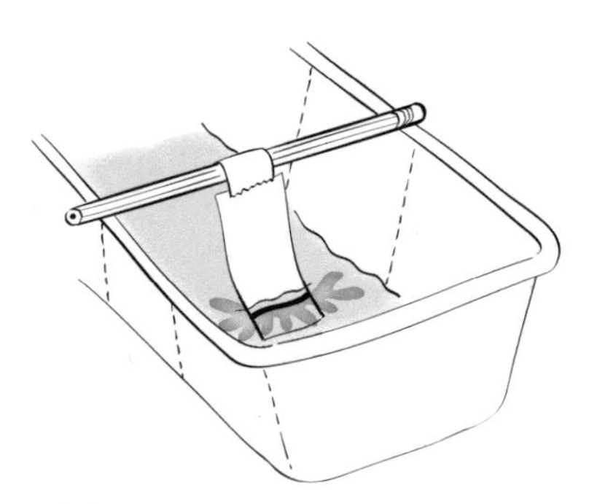

If the pen mark is put below the water level, the ink will leach out into the liquid instead of traveling up the paper.

5. Distribute a sheet of scratch paper and a piece of the ransom note to each *pair* of students, and have them conduct a chromatography test on their piece of the note.

Some students are convinced that the ink marks are traveling downward, rather than upward. Encourage careful observation to notice that the ink is carried upward by the water.

6. After the students have conducted their tests, reconvene the class. Have the students share some of their observations. Again, encourage students to participate and ask questions to help them be as precise as possible in their observations.

7. Explain to the students that:

   - The colors they see are pigments that were mixed together to make black ink. A ***pigment*** is a chemical that makes something look a certain color. Ink is a mixture of different substances. Different companies make black ink with different combinations of pigments.

   - Chromatography is a technique that separates ***substances*** from a ***mixture.*** Each different mixture will produce its own color pattern (called a ***chromatogram).***

Some teachers like to dissect the word chromatography here. Ask if students can think of any other words that have "graph" or "graphy" in them. "Chromo" means color, and "graph(y)" means writing.

8. If you've decided to use them, distribute the magnifying lenses now. They may help students detect the presence of any faint pigments.

9. Ask a student to describe the color at the very bottom of the ransom note chromatogram. Ask if anyone disagrees and encourage them to share their results. Ask for raised hands if there seems to be disagreement. Follow this same procedure with each of the colors in the chromatogram. Your students will probably disagree on some details, but agree on the general pattern of colors.

Vocabulary*. As new vocabulary is introduced in the context of this activity and later ones, you may want to post some of the new words and definitions on the wall for students to refer to throughout the unit.*

10. If there is disagreement about the colors, ask why different groups might have different results. [The amount of ink applied to the

paper, length of time paper was left in water, that different people perceive colors differently, etc.] Introduce the term ***variable*** as something that can change from one test to the next.

■ Introducing the Suspects

1. Post or reveal the list of suspects. Ask for volunteers to read the parts of the suspects, then choose those who are comfortable reading in front of the class. Tell the class that in this case their classmates are like actors portraying the characters, not the characters themselves.

2. Pass out one suspect statement to each student actor. Remind them that—when their turn comes—they should face the audience and read in a loud clear voice—and not too fast. Read the "teacher" lines from the "Introducing the Suspects" section of the **Teacher Script,** as the students each read their assigned parts.

3. As each suspect is introduced, point out the name on the list, and the number of the pen.

■ Testing the Mystery Pens

1. Ask your students to briefly discuss with their table group or partner how they could use chromatography to discover which suspect's pen was used to write the ransom note. Ask a group to share their idea with the class, and as needed make sure the basic idea is clear. [They could test the ink in each pen and compare the results with the ransom note test results.] This may seem obvious, but students often need to verbalize the process.

2. Before setting students to work, emphasize the following important procedural points:

a. Each pair of students will get blank paper strips to test all six pens. Each pair should work as partners and share the tasks.

b. Point out the six stations to the students and say they'll go to the pen stations to put the pen marks on each test strip. They will draw a horizontal line across the paper strip about an inch up from the bottom. Making the mark at this height will help keep the pen mark out of the water. They should leave the pen at the station.

c. Let students know that while they're still at the station they will use a *pencil* to number the test strips, writing the appropriate pen

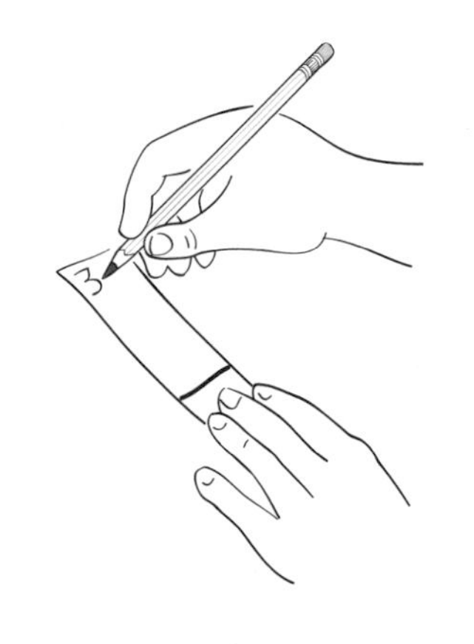

number near the *top* of each strip. Point out that labeling them in ink might cause smearing and or writing near the bottom of the strip could interfere with the test.

d. Tell students they should do the rest of the tasks back at their table to avoid crowding at the pen stations.

e. They should tape the strips to pencils. Let them know they will probably need to use two pencils, since not all six strips will fit into the trough on only one pencil.

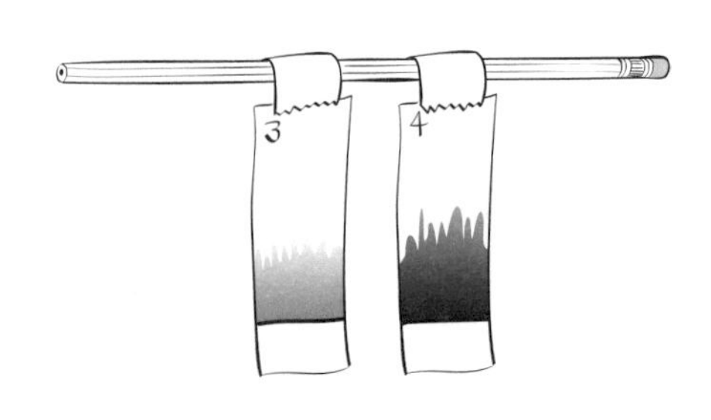

f. When the water has traveled about three-quarters of the way up the strip, remove it from the trough and tape it to the sheet of scratch paper *next to the ransom note strip.* Ask students to write their names on the scratch paper.

3. Have students begin. Circulate and remind students of the procedure as necessary.

4. When everyone has finished the chromatography tests, have student volunteers collect the materials, including the papers with chromatograms, which can be laid out to dry.

5. Tell students that, in the next class session, you'll ask them for their chromatography test results. Mention that some evidence might not be visible when the test strip is wet, and sometimes more can be seen after it dries.

Session 2: Solving the Mystery

■ What You Need

For the class:

- ❑ list of suspects from the previous session
- ❑ 1 overhead transparency of **Different Water Molecule Models** (page 27)
- ❑ an overhead projector

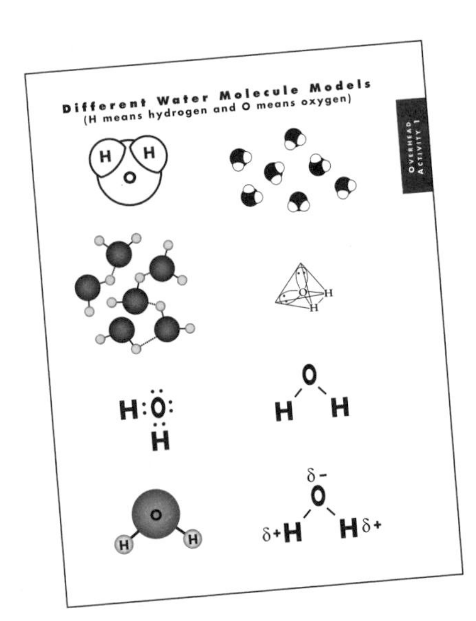

For each pair of students:

- ❑ their scratch paper with chromatograms from the previous session
- ❑ *(optional)* 1 magnifying lens

For each student:

- ❑ 1 piece of unlined paper to draw on
- ❑ 1 pencil

■ Getting Ready

1. Make an overhead of **Different Water Molecule Models** (page 27).

2. Have on hand the chromatograms that each pair of students made in the previous session as well as the list of suspects.

3. Have the unlined drawing paper and pencils ready to distribute during the "Microscope Eyes" Model part of the session.

4. If you've decided to use them, set out the magnifying lenses.

■ Identifying the Mystery Pen

1. Pass out their papers with the chromatograms and, if you're using them, the magnifying lenses. Ask the student pairs to spend the next few minutes carefully examining their chromatograms to determine which pen they think was used to write the ransom note.

2. Encourage the students to examine another group's chromatograms to see if the two sets of evidence are consistent.

3. Reconvene the class. Ask your students if they think they know which is the "mystery pen." Have them present their evidence, comparing the colors in the chromatogram of the ransom note with the colors in the chromatograms of the pens of the suspects.

4. Continue this discussion until your class reaches agreement on which pen matched. This usually happens fairly quickly. Refer to the list of suspects as needed.

■ Does the Evidence Prove the Suspect is Guilty?

1. Ask students to raise their hands if they think they have enough evidence to convict the suspect whose pen matches the one used for the ransom note. [Most will probably raise their hands.]

2. Ask who thinks the matching pen is *not* enough evidence to convict the suspect. [A few hands will probably go up.]

3. Acknowledge that one explanation for the evidence is that the suspect wrote the note with his/her pen, and is therefore guilty. But challenge your students to act as defense attorneys and try to come up with explanations of how the suspect could be innocent, even though his/her pen matches the mark from the note.

4. Have student pairs discuss this for several minutes to refine their thinking. Remind them that they need to be able to explain their reasoning.

5. Reconvene the entire class and ask a number of students to state and briefly explain their ideas as defense attorneys. Students often say:

   - The suspect was framed. The pen was borrowed by someone else who committed the crime; that person tried to make it look like the suspect was guilty by writing the note with his/her pen.

   - Someone else had the same brand of pen and wrote the note. The suspect is not the only person in the world who owns a pen of that brand!

6. After a thorough discussion, ask the students to vote again on whether or not they think the matching pen is enough evidence to convict the suspect. [Hopefully you will have many more hands now voting that it is **not** enough evidence.]

7. Point out that in the real world, more evidence would be needed to convict someone.

■ The "Microscope Eyes" Model of Chromatography

One good way for students to share their answers to your questions is to hold up the number of fingers that matches the number of the pen in question.

1. Have the students refocus on their chromatograms. Ask for observations about how the chromatograms compare, including:

   - Which ink traveled the farthest up their test strips?

   - Which ink contained the greatest number of different colors or pigments?

   - Were there any inks that did not travel up the paper or separate into other colors?

2. Ask for their ideas on why some pigments (colors) traveled higher than others.

3. Tell the class that all substances, including water, ink pigments, and paper are made up of atoms and molecules. ***Molecules*** are extremely tiny particles. Different substances have molecules of different shapes, sizes, and weights.

4. Draw the classic "Mickey Mouse" version of a water molecule on the board. Tell them a water molecule consists of two hydrogen atoms bonded with one oxygen atom.

5. Tell them that although we can't see water molecules, scientists draw them like this based on their evidence of how water behaves. Although this may not be what real water molecules look like, it is a ***model*** that helps scientists think about the tiny molecules and why they behave the way they do.

You may also want to hold up a different example of a model, such as a toy car or doll, and ask your students what is accurate and what is inaccurate about this model. How might they change it to make it more accurate?

6. Show the **Different Water Molecule Models** transparency. Let students know that each of these models has strengths and weaknesses, showing some things well, and other things not as well. Point out that all models are inaccurate in some way, because to be *completely* accurate they'd have to be the real thing.

7. Ask students to pretend that they have "microscope eyes" and can see what pen ink molecules, paper molecules, and water molecules look like.

8. Say they are going to draw a picture or "model" of how they imagine the molecules might be shaped. Their drawing should try to show why some ink molecules go up further than others on the paper. Tell them this drawing is just a first draft, and they will be given a chance to redraw it later. Emphasize that they don't need to be "correct." Let them know that starting to draw this model is a way for them to begin to express their own ideas about how the shapes, interactions, or other characteristics of the water, ink, and paper molecules might explain why pigments are carried up the paper and why some are carried higher than others.

9. Pass out unlined paper and pencils to each student. Ask students to write their names on the paper and label it as "Microscope Eyes Model of Chromatography." Then have them begin drawing.

10. Circulate as they draw, and encourage them to jot down notes about what parts of chromatography or their model they are un-

sure of. Stimulate their thinking with questions such as "Is there something that doesn't quite make sense?" "What does your model explain well?"

11. A few minutes before the end of the session, ask some of the students to share their drawings with the class. Reinforce the idea that what they have drawn is a type of model. It *represents* something, but is not the actual thing. Either collect their drawings, or tell students to hold on to them, so they can make adjustments to them in the next activity.

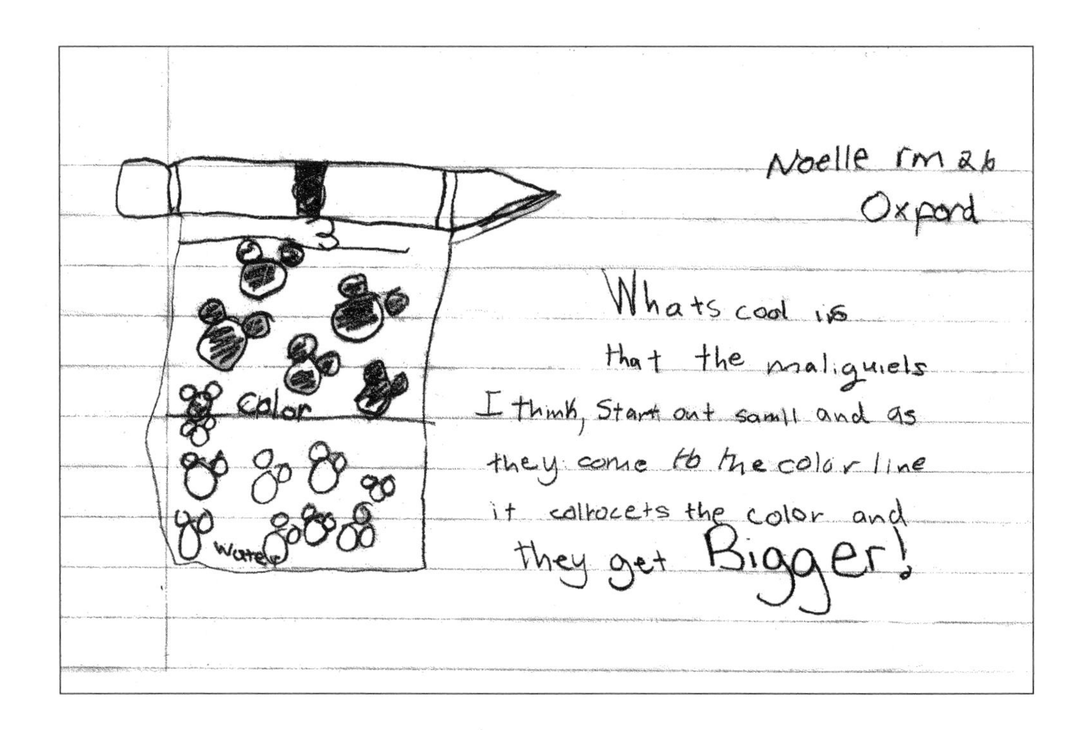

TEACHER SCRIPT
The Mystery of the Missing Cold Remedy

Introducing the Mystery

Be sure your students know this is a make-believe mystery!

Teacher: You're not going to believe this, but on my way here this morning, I was stopped by a police officer. Apparently there was a theft at a nearby school. The school held a science fair a couple of months ago, and one of the student projects was called "Solution X, the Cure for the Common Cold." It generated a lot of interest at the fair, and at the end of the fair, Robin Yall, a parent of a student at the school, was caught trying to steal the project. Ms. Yall has since moved to another state.

The project was put in the school safe, but this morning the safe was found open, and the science fair project was missing. In its place was a note written with a black felt-tip pen. The note said:

"You can catch a cold, but you can't catch me,
I'll run like a nose, unless I get a fee.
Number 1 with six zeros after.
Pardon while I launch into maniacal laughter.
Bwa ha, ha, ha, ha, ha, ha, ha, ha, ha, ha, ha, ha!"

We will hear from the suspects a little later, but first let's use chromatography to test the ink mark made by the black pen.

Introducing the Suspects

Teacher: The police officer told me they have six suspects, and I'd like to introduce you to them. The first suspect is Kat Chacold, a parent of another student at the school.

Have a student volunteer read each suspect statement.

1. Kat Chacold: I work for a company that makes medicine. I'm not a common thief!

Teacher: Perhaps. But we do know that during the science fair Ms. Chacold was asking the student who did the project a lot of questions. What if she stole the recipe for her company? A cure for the common cold that really worked could be worth a fortune!

Our second suspect is Ed Chew.

2. Ed Chew: I teach home economics at the school. I'm an award-winning cake baker. I have no interest in cold remedies.

Teacher: Ed Chew is really muscular, and so are a lot of the kids in his class. We've heard that he got angry because they took away his electric mixers due to budget cuts. There's even a rumor that he gives steroids to the kids in his class to build up their muscles so they can make cakes that are light and fluffy. What if Ed Chew stole the project to sell, and make money to buy equipment for his class?

The next suspect is the principal, Ronnie Noze.

3. Ronnie Noze: I'm the principal and the coach of the local champion softball team. I am an upstanding member of the community, not a thief.

Teacher: That's true, and it's also true that Mr. Noze has the best pitcher on his team, who always seems to have a runny nose. She catches a lot of colds, but never misses a game. Some people wondered if her runny nose has something to do with why the balls she pitches are so hard to hit. Maybe Ronnie Noze stole the project to prevent the cure for the common cold from reaching shelves, because he's afraid his pitcher wouldn't be as good if she didn't always have a cold.

The next suspect is Ivana Tishu, another parent.

4. Ivana Tishu: I'm a mother with a big family. I've dedicated my life to my children, not to stealing.

Teacher: We don't know much about her, except that her six kids are always sick, and their family has to get the big size garbage cans, just to hold all the dirty tissues they go through. Maybe she stole the cure to help out her own family…and the local landfills.

Our next suspect, is Noah Snotmee.

5. Noah Snotmee: My daughter goes to the school that was robbed, but I don't work there, and I have had nothing to do with it. It's not me you're looking for. It's not me!

Teacher: We've heard Noah was very angry because his daughter's science project, "Cosmetics for Cats," was out shone by the "Solution X, the Cure for the Common Cold" project. He worked very hard on the project. Oh yeah, and I think his daughter helped too. So perhaps he's angry about his daughter's project being rejected, and decided to get revenge by stealing the "Cure for the Common Cold" project.

Our next suspect is Hannah Kercheef.

6. Hannah Kercheef: I'm a successful clothing designer. But I didn't steal your recipe. I don't need it.

Teacher: Hannah Kercheef has a clothing company. She has made a lot of money selling shirts she designed with super absorbent material on the sleeves. They are called "hanky sleeves" and they are for wiping your nose. Hmm. Maybe Ms. Kercheef is afraid that if the common cold is cured, she'll be out of business.

SUSPECT STATEMENTS

The Mystery of the Missing Cold Remedy

1

Kat Chacold

I work for a company that makes medicine. I'm not a common thief!

2

Ed Chew

I teach home economics at the school. I'm an award-winning cake baker. I have no interest in cold remedies.

3

Ronnie Noze

I'm the principal and the coach of the local champion softball team. I am an upstanding member of the community, not a thief.

4

Ivana Tishu

I'm a mother with a big family. I've dedicated my life to my children, not to stealing.

5

Noah Snotmee

My daughter goes to the school that was robbed, but I don't work there, and I have had nothing to do with it. It's not me you're looking for. It's not me!

6

Hannah Kercheef

I'm a successful clothing designer. But I didn't steal your recipe. I don't need it.

TEACHER SCRIPT

The Mystery of the Missing Dinosaur

Introducing the Mystery

Be sure your students know this is a make-believe mystery!

Teacher: The local science museum has a mechanical dinosaur—or should I say *had* a mechanical dinosaur. Yesterday when the museum was closing, they discovered that the mechanical dinosaur was missing. On the floor where the dinosaur once stood was a note. The note was written with a black felt-tip pen. The note said:

> *"Ha, ha, ha, ha. I'm smarter than all of you and you'll never catch me! If you ever want to see your dinosaur again, you will give me one million chocolate bars."*

We will hear from the suspects a little later, but first let's use chromatography to test the ink mark made by the black pen.

Introducing the Suspects

Teacher: There were not very many people in the museum at that time, but six of them each had a black felt pen on their persons. They are all suspects. Here is what they have to say:

Have a student volunteer read each suspect statement.

1. Cara Lott: I am a mother of three children and I lift weights. I work for a group that helps dying children get their last wish. I would never steal a penny.

Teacher: Cara seems like a very kind and giving person, but maybe she's kind enough to *steal* the dinosaur! It is rumored that one of the dying children she works with wished for a dinosaur of his own! Did she steal the dinosaur to help fulfill a dying child's dream?

2. Ben Alteebocks: I got most of my teeth knocked out when I was a hockey player. I was a pro athlete and I made a lot of money. Why would I steal?

Teacher: Did Ben Alteebocks really lose his teeth playing hockey? Or did too much sugar rot them? It might also have something to do with his being a CHOCOHOLIC! We found his name on the register at a chocoholic institute. Could he have kidnapped the dinosaur to feed his habit with candy bars?

3. Trudy Lenz: You may have seen some of my work on TV. I make music videos. I went to the science museum to see the exhibits and eat at the cafe.

Teacher: We've found out that Trudy's last music video was "Rock the Dinosaur." We wonder if Trudy may have taken the dinosaur to dance in one of her videos.

4. Moe Studly: Yo dudes, I am a handsome college student. I work part time at the science museum. I'm very popular and buff. A foxy hunk like me wouldn't steal a dinosaur!!!!!!

Teacher: Moe is very popular, and especially popular with... Trudy!... because she is his girlfriend. People say that Trudy has Moe wrapped around her finger, and that he would do anything she told him to do, perhaps even including steal a dinosaur.

5. Professor Terry Daktil: I teach my students about dinosaurs and went to look at the exhibit. I am not foolish enough to steal a dinosaur.

Teacher: We also happen to know that the professor, Terry Daktil, is putting together his own little exhibit of dinosaurs for his students. Could he be taking from the museum's exhibit to build his own?

6. Vera Klooless: What dinosaur? I didn't even know there were dinosaurs at the museum. I don't like science, and only went there because my brother made me.

Teacher: Vera may know more than she's letting on. When we found her in the museum, she had a copy of a high level book about dinosaurs with her.

SUSPECT STATEMENTS
The Mystery of the Missing Dinosaur

1

Cara Lott

I am a mother of three children and I lift weights. I work for a group that helps dying children get their last wish. I would never steal a penny.

2

Ben Alteebocks

I got most of my teeth knocked out when I was a hockey player. I was a pro athlete and I made a lot of money. Why would I steal?

3

Trudy Lenz

You may have seen some of my work on TV. I make music videos. I went to the science museum to see the exhibits and eat at the cafe.

4

Moe Studly

Yo dudes, I am a handsome college student. I work part time at the science museum. I'm very popular and buff. A foxy hunk like me wouldn't steal a dinosaur!!!!!!

5

Professor Terry Daktil

I teach my students about dinosaurs and went to look at the exhibit. I am not foolish enough to steal a dinosaur.

6

Vera Klooless

What dinosaur? I didn't even know there were dinosaurs at the museum. I don't like science, and only went there because my brother made me.

Different Water Molecule Models
(H means hydrogen and O means oxygen)

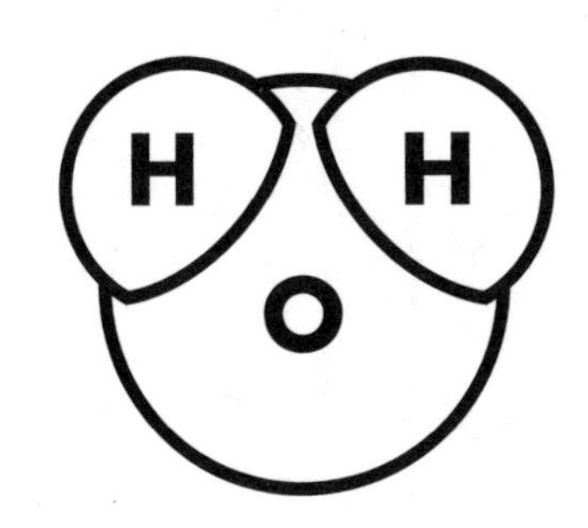

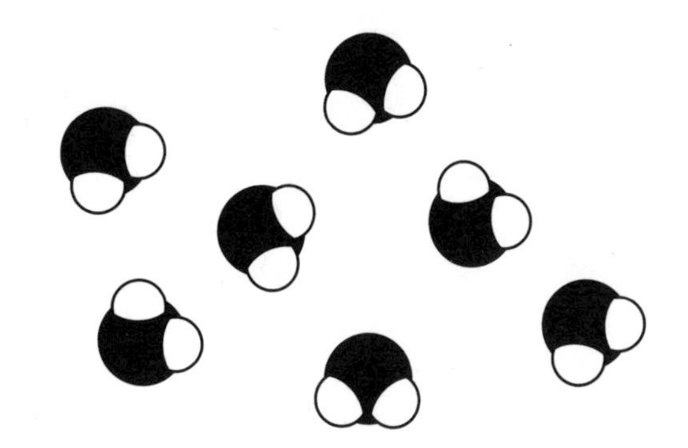

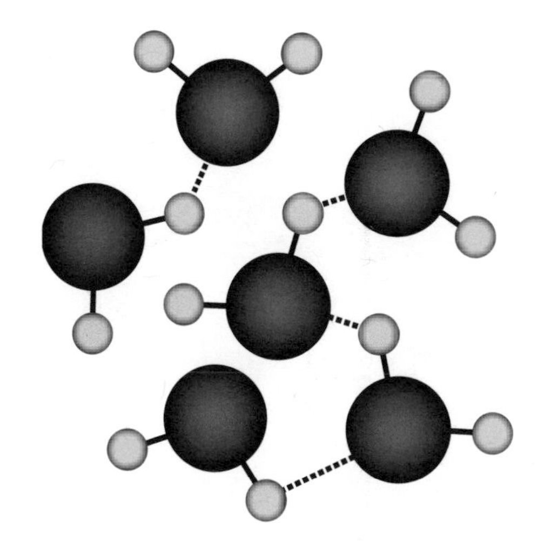

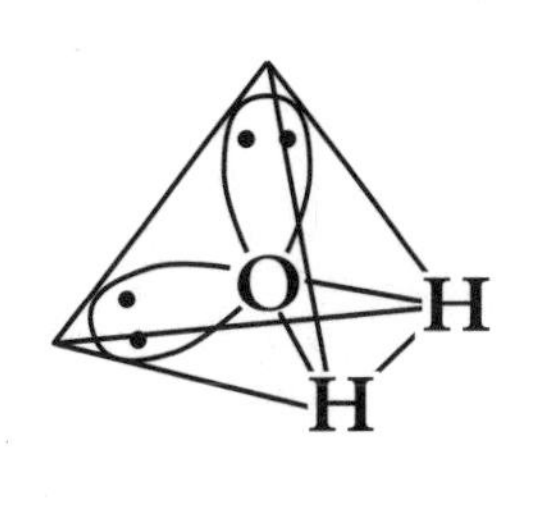

H:O:
H

O
H H

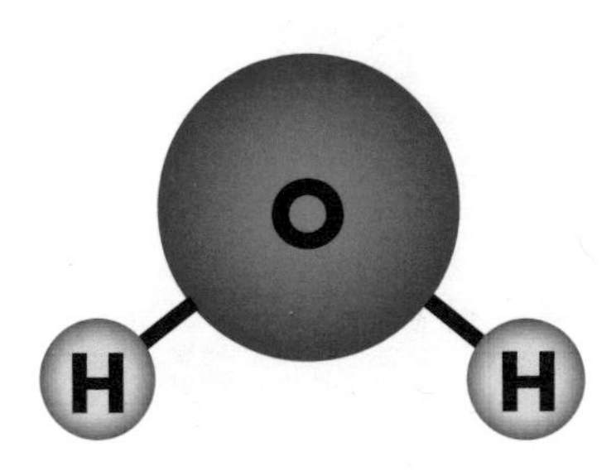

δ-
O
δ+H H δ+

Overview

In this activity, students learn that paper chromatography is only one of many types of chromatography used in diverse fields of science. They also learn that all forms of chromatography can be viewed as a *system,* composed of a medium, a solvent, and a test substance.

The teacher then gathers the class around a large chromatography model, in which the molecules of the test substance are represented by rocks, paper wads, styrofoam balls, and styrofoam packing "peanuts." The medium in the model is represented by the floor and the solvent is the moving air from a fan or hair dryer. After seeing the demonstration, the students discuss how they could change the medium and the solvent in the model to separate all the molecules of the test substance more effectively.

Next, students examine four different models of the ink-paper-water chromatography system they used in Activity 1, and learn some more facts about molecules. They are then asked to consider what is accurate and inaccurate about each of the four models. Finally, either in class or as homework, students are challenged to redraw their original molecular models to include the new information they have learned about molecules and chromatography. This second drawing can serve as an assessment of students' understanding.

Learning Objectives for Activity 2

- Help students understand that chromatography is a **system** composed of a medium, a solvent, and a test substance.
- Have students understand that scientists use chromatography in many ways, and that mediums and solvents are selected based on how well they serve to separate the parts of the test substance.
- Enable students to learn more about molecules and how their shape, size, and weight affects their interactions with different substances.
- Have students learn to revise their own models as scientists do—based on new information and evidence.

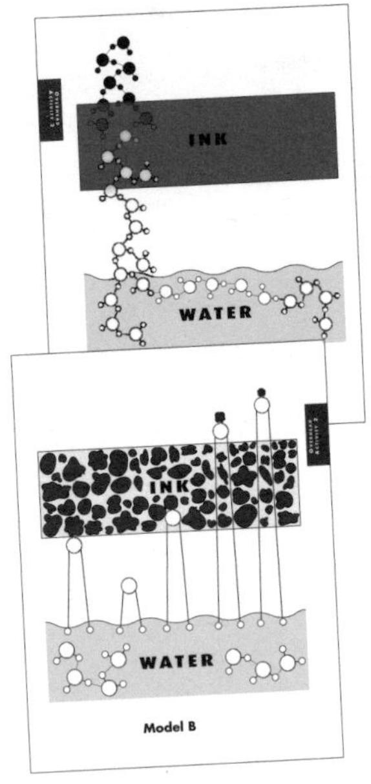

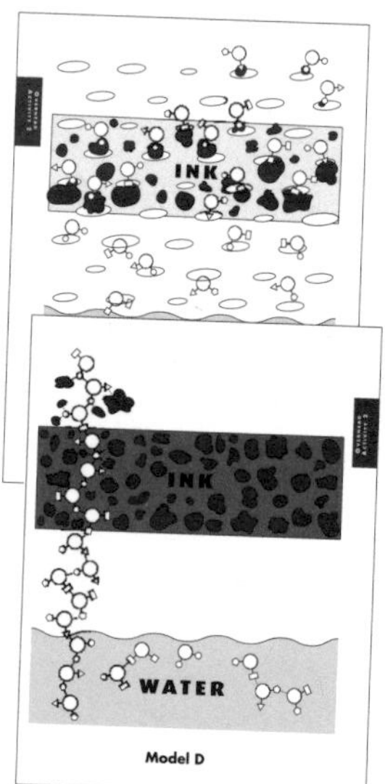

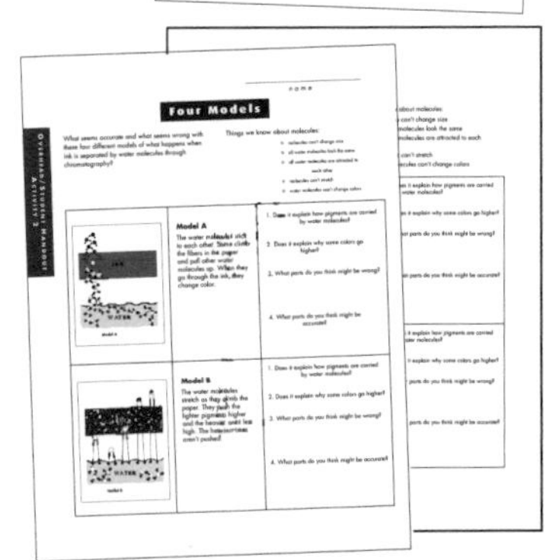

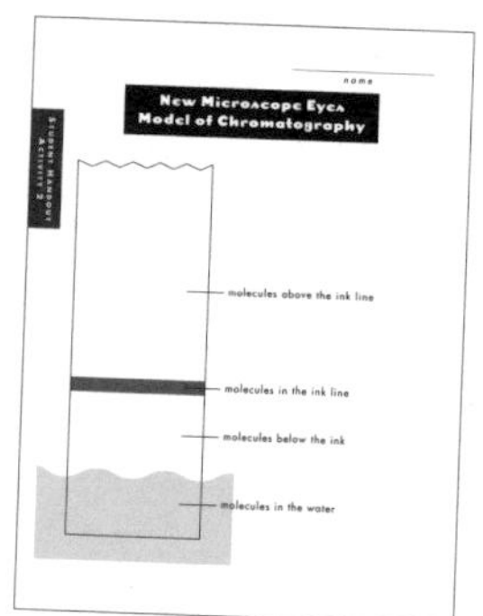

■ What You Need

For the class:

- ❑ about 5 small rocks, any kind, from about 1"–3"
- ❑ about 5 wadded up individual sheets of $8\frac{1}{2}$" x 11" paper
- ❑ about 5 styrofoam balls, about 2" in diameter
- ❑ about 5 styrofoam packing "peanuts" (or other lightweight items)
- ❑ a large fan or blow dryer
- ❑ 1 overhead transparency of each of the following:
  - __ **Model A** (page 38)
  - __ **Model B** (page 39)
  - __ **Model C** (page 40)
  - __ **Model D** (page 41)
  - __ the **Four Models** student sheet (pages 42–43)
- ❑ an overhead projector
- ❑ *(optional)* an extension cord for the fan or blow dryer

For each student:

- ❑ their "Microscope Eyes Model of Chromatography" drawing from the previous activity
- ❑ 1 copy of the **Four Models** student sheet (pages 42–43)
- ❑ 1 copy of the **New Microscope Eyes Model of Chromatography** student sheet (page 44)
- ❑ a pencil
- ❑ *(optional)* a blank sheet of $8\frac{1}{2}$" x 11" paper for the "Questions" sheet used at end of activity

■ Getting Ready

1. Choose a long section of *non-carpeted* floor for your demonstration of a model chromatography system. (See the description on pages 32–33.) The section of floor should be long enough for all of your students to stand in two lines facing each other, creating a "lane" about three feet wide between the two lines of students. If your classroom is too crowded or is carpeted, plan to use a hallway or other space for this short demonstration.

2. Plug in the fan or blow dryer at one end of the section of floor.

3. Wad up about five sheets of $8\frac{1}{2}$" x 11" paper. Gather about five styrofoam balls, five styrofoam "peanuts," and five small rocks. Have them handy near where you will do the demonstration.

4. Turn on the fan or blow dryer and test the model before class to make sure there is some separation of the objects by the blowing air. Ideally, some of the objects will be moved at least several feet along the floor, and others will be moved a shorter distance.

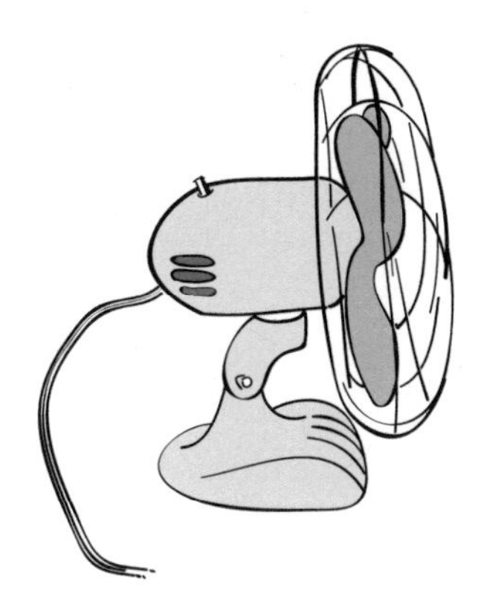

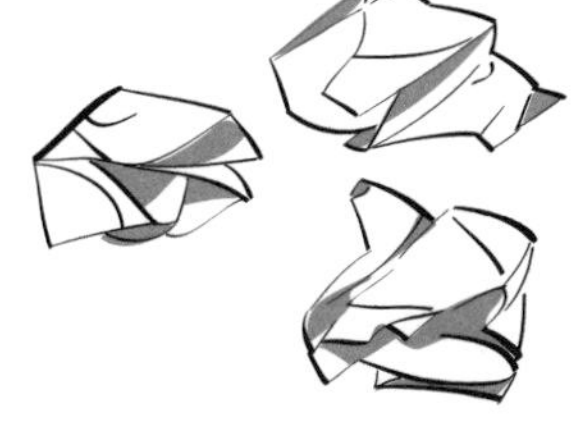

If some of the objects blow off to the side, you might want to set broomsticks, yardsticks, or boards on either side of the section of floor to keep them in a "lane."

5. Make one overhead of each of the following: **Model A, Model B, Model C,** and **Model D** (pages 38–41), and **Four Models** (pages 42–43).

6. Make one copy each of the **Four Models** student sheet (pages 42–43) and the **New Microscope Eyes Model of Chromatography** student sheet (page 44) for each student.

7. Have a pencil and their "Microscope Eyes Model of Chromatography" drawing ready to distribute to each student. If you plan to do the optional "Sharing the New Microscope Eyes Models" at the end of the activity, have a blank sheet of paper for each student.

8. During the "Revisiting the 'Microscope Eyes' Model" part of the activity, your students will work in discussion groups of about four to six students. If necessary, plan ahead how to re-group students.

■ Explaining the Chromatography System

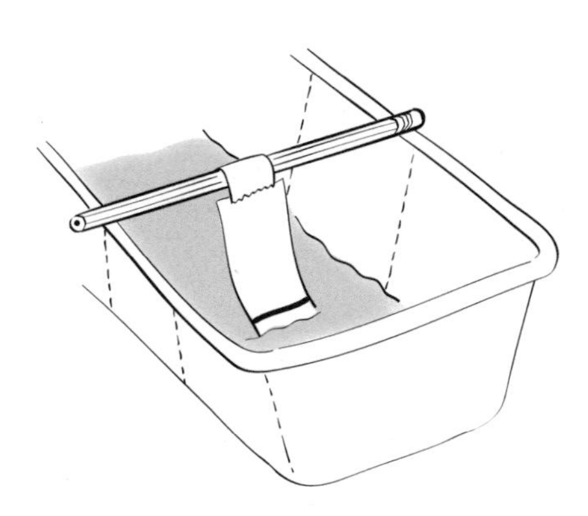

1. On the board or overhead, sketch the chromatography set up used in Activity 1 (a strip of paper hanging from a pencil or straw, a pen mark, and water in a trough or cup).

2. Remind the students that chromatography is a technique used by many different kinds of chemists to separate mixtures into their parts. Say that to understand how chromatography works, it's helpful to analyze how chromatography is a system.

3. Define a ***system*** as a group of interacting parts that work together. Tell students that in the chromatography system there are always three main parts. Label them on your drawing as you introduce them:

   a. Ask, "What was the substance you were testing?" [The ink.] Tell them this is called the ***test substance.***

   b. In chromatography, the ***medium*** is what the test substance moves through. Ask, "What was the medium in your chromatography system?" [The paper.] Explain that this is why it's called *paper chromatography.*

   c. The third part of the chromatography system is the ***solvent.*** It is what helps move the test substance through the medium. Ask, "What was the solvent in your system?" [Water.]

This is a great opportunity to make the connection for your students to the big idea of systems that permeates all disciplines of science. For a fuller definition of a system and more information about chromatography in general, please see "Background for the Teacher" on page 65.

4. Ask why the ink from one of the pens didn't travel up the paper at all. [It's a "permanent" pen. Water didn't work as a solvent for this ink.] Ask, "Do you think another solvent would make the permanent ink molecules travel up the paper?" [If no one mentions it, suggest that alcohol or vinegar might work as a solvent for this pen.]

5. Explain that there are many types of solvents and mediums used in chromatography, depending on the test substance you want to separate. Say that chromatography is an incredibly important process and tool in chemistry, because it can be used to separate the parts of almost any substance.

■ Demonstrating a Model Chromatography System

1. Gather everyone around the floor area you have chosen for your model. Say you're going to use a large model to demonstrate how chromatography works. Remind them that all models have inaccuracies, but models help us understand things that are hard to see directly.

2. Ask the students to imagine that they are chemists who want to separate the parts of a test substance. Set the styrofoam balls, styrofoam "peanuts," rocks, and wads of paper in a pile in front of the fan or blow dryer. Tell students all these objects will represent the test substance—a mixture of different molecules of different shapes and sizes. Emphasize that molecules from different substances are different from each other.

3. Say the floor will represent the medium in this model. Point out the fan. Tell them the moving air will represent the solvent.

4. Ask students to predict what will happen to the test substance once the air is blowing and the model is in operation.

Like all models, this one has its limitations. It demonstrates well how the substances move through the medium due to differences in their mass (with some connection to shape and size), but it doesn't address the issue of attraction of the substances to the medium nor the solubility of the substances.

5. Caution the students to resist touching the objects. Put the fan on a low setting, and ask students what they observe. [Usually, the styrofoam travels farthest, the paper wads less far, and the rocks don't move at all.] Ask why they think it is happening.

6. Gradually increase the amount of moving air to make reluctant objects move along. You may need to shift the direction of the air somewhat to move along wads of paper or styrofoam that are caught behind a rock.

7. When you have achieved some pattern of separation, turn off the fan. Ask what made some objects in this model travel farther than others. [Objects had different sizes, weights, or shapes, or different attraction to the medium.]

8. Tell them that these are all reasons real substances separate from each other in real chromatography. In this model, the rocks represent a substance that is not soluble in the chosen solvent, like the ink from the permanent pen.

9. Ask, "Which objects didn't separate well in this model?" [Usually, the styrofoam peanuts and styrofoam balls remain mixed with each other.] Ask, "How could we adjust the system so that these objects separate?" [Make the medium a carpet or fabric; increase the amount of moving air.]

10. Have all students return to their seats.

■ Revisiting the "Microscope Eyes" Model

1. Refer back to your drawing to briefly summarize how the paper chromatography system works, integrating the ideas offered by the students in previous activities into your explanation as much as possible:

> As the water passes upward through the ink, some pigments in the ink mixture are attracted to the water and are carried up the paper by it. Some pigments are more attracted to the water than others, so they travel farther up the paper. The distance a pigment travels up the paper depends on a number of factors, including (a) what solvent is used, (b) the weight, shape, and size of the pigment molecules, and (c) how strongly attracted the pigment molecules are to the medium and the solvent.

2. Say that, now that they know more about chromatography, they'll have a chance to analyze four drawings (which were not made by anyone in the class) that are possible models for how the ink molecules separated in the chromatography test. **Emphasize that all four models have some things that are accurate and some things that are not accurate.**

3. Display the transparencies of chromatography **Models A** through **D** on the overhead projector one at a time, and *briefly* explain each model. Mention that in all four models, the solid lines in the molecules imply attachment and the dashed lines between molecules imply attraction.

Model A
Water molecules are attracted to each other and to fibers in the paper. These attractions cause them to move up the paper. When they go through the ink, they change color.

Model B
The water molecules stretch as they climb the paper. The lighter weight pigments move higher and the heavier ones less high. The heaviest ones don't move up.

Model C
The water molecules move up through tiny holes in the paper. The ink pigments are attracted to the water molecules. The bigger pigments can't fit through some holes so they don't go as high.

Model D
The pigment molecules are attracted to the water molecules and to some extent to the paper molecules. The pigments that are more attracted to paper than to water don't go as high. The pigments more attracted to water than paper go higher.

4. Put the transparency of the **Four Models** student sheet on the overhead. Say that they will each get a sheet like this.

5. Explain that after many years of scientific investigation, scientists have learned some facts about molecules that may help students figure out what is accurate or inaccurate about these models. Encourage students to re-read these facts as they think about each model. Go over the "Things we know about molecules" at the top of the sheet:

- molecules can't change size

Molecular Disclaimer
These "rules" about molecules are necessarily oversimplified to make this activity accessible to students and convey a basic sense of what is happening in the chromatography process. For this purpose, they are helpful. Especially because the students are only considering water molecules and their behavior during chromatography, the statements are sufficiently accurate. From a more advanced scientific perspective, some of the "rules" would need to be qualified and/or more fully detailed. It is true, for example, that the water molecules the students are considering do not stretch or change size—they definitely do not expand a lot or stretch up the paper the way Model B suggests. However, this is not necessarily true of all molecules, as under certain conditions some can slightly change in size and/or stretch/bend. At a higher level of scientific sophistication, numerous provisos to these "rules" would need to be made.

- all water molecules look the same
- all water molecules are attracted to each other
- molecules can't stretch
- water molecules can't change colors

6. Give a **Four Models** sheet to each student. Tell them they should first read the pages, then discuss in their small group what is inaccurate or accurate about each model. Then each student should write answers on her own sheet.

7. After students have discussed and recorded their answers, regain the attention of the class. Hold a brief discussion of what is accurate or inaccurate about each model. Based on their prior experiences and the facts about molecules listed at the top of their sheets, students should bring out most of the following points:

Model A
Accurate: Water molecules are attracted to each other and to paper so they do move up the paper.
Inaccurate: The water molecules don't change color when they pass through the ink.

Model B
Accurate: Lighter weight pigments move higher simply because they're lighter.
Inaccurate: Molecules can't really stretch.

Model C
Accurate: Water molecules move up the paper through tiny holes. The ink pigments are attracted to the water molecules. Bigger pigments can't fit through the holes.
Inaccurate: Water molecules don't look different from each other. They should all look the same.

Model D
Accurate: Pigment molecules are attracted to water and paper molecules. Pigments that are more attracted to the paper don't go as high. Pigments that are more attracted to the water go higher.
Inaccurate: Water molecules don't look different from each other. They should all look the same.

Advantages and Limitations of Models
Emphasizing both what is "accurate" and "not accurate" about the models should help students recognize both the strengths and limitations of a particular model. Only focusing on what is "wrong" or what is counter to the molecular "rules" can limit the recognition of a model's strengths. Given this, some teachers also frame the discussion of each model by posing the question as: "What does the model explain or not explain?" so the discussion focuses on the advantages and limitations of the models, rather than focusing only on "what is wrong" about a particular model. This also helps students think about the advantages and limitations of models in general, not in terms of whether they are "right" or "wrong," but how well they help explain or clarify a particular process.

■ Revising Their Earlier Models

If time is short, you may decide to make all or part of this activity a homework assignment. These drawings can serve as a great assessment of students' understanding.

1. Review that they now know what scientists think causes the separation of molecules in a chromatography test: It has to do with the size, weight, and shape of the molecules and how strongly they are attracted to the solvent or the medium. Go on to say that scientists don't have the perfect model of what goes on at the molecular level, and there is always room for improvements to the scientific model.

2. Say that one sign of a good scientist is the ability to change your ideas when presented with new evidence that conflicts with earlier ideas. Tell them their next challenge is to make a new "microscope eyes" drawing of the chromatography test of the ink, incorporating what they now know about the chromatography system and molecules.

3. Say that you will hand back the "Microscope Eyes Model of Chromatography" drawings they did in Activity 1 so they can see if their ideas have changed. Tell them their new drawing can be similar to the one they did before, or completely different. Remind them that it need not be pretty.

4. Hold up the **New Microscope Eyes Model of Chromatography** student sheet. Tell students they'll draw what they think the molecules might look like in the appropriate places on the sheet.

5. Say their model should show their current idea of what the different ink pigments look like and explain why some of them go higher on the paper than others. Mention that they can refer to their earlier drawing and the **Four Models** sheets.

6. Give students their first drawing, then pass out the **New Microscope Eyes Model of Chromatography** student sheet. Have students make their new drawings.

■ (Optional) Sharing the New Microscope Eyes Models

1. Assign students to groups of four. Give each student a pencil and a sheet of paper to be labeled "Questions." Tell them to write their name on their Questions sheet, as well as on their **New Microscope Eyes Model of Chromatography** drawing.

2. Tell each student to pass their two papers (Questions and **New Microscope Eyes Model of Chromatography**) to the person sitting to their left in their group. Tell them to study the drawing handed to them, try to figure it out, and write any questions they have about it on the accompanying Questions sheet.

3. After a few minutes, tell them to pass their papers to the left again and follow the same procedure. Continue this until every student receives his or her own papers again.

4. Give students a few minutes to read through the questions written by their colleagues and think about them. Tell them to think of how they might want to readjust their drawing after reading the questions.

5. Tell the groups to take turns discussing the drawings of each of their members one at a time. They can describe each model, ask questions (those they wrote or new ones), and respond to the questions of their colleagues.

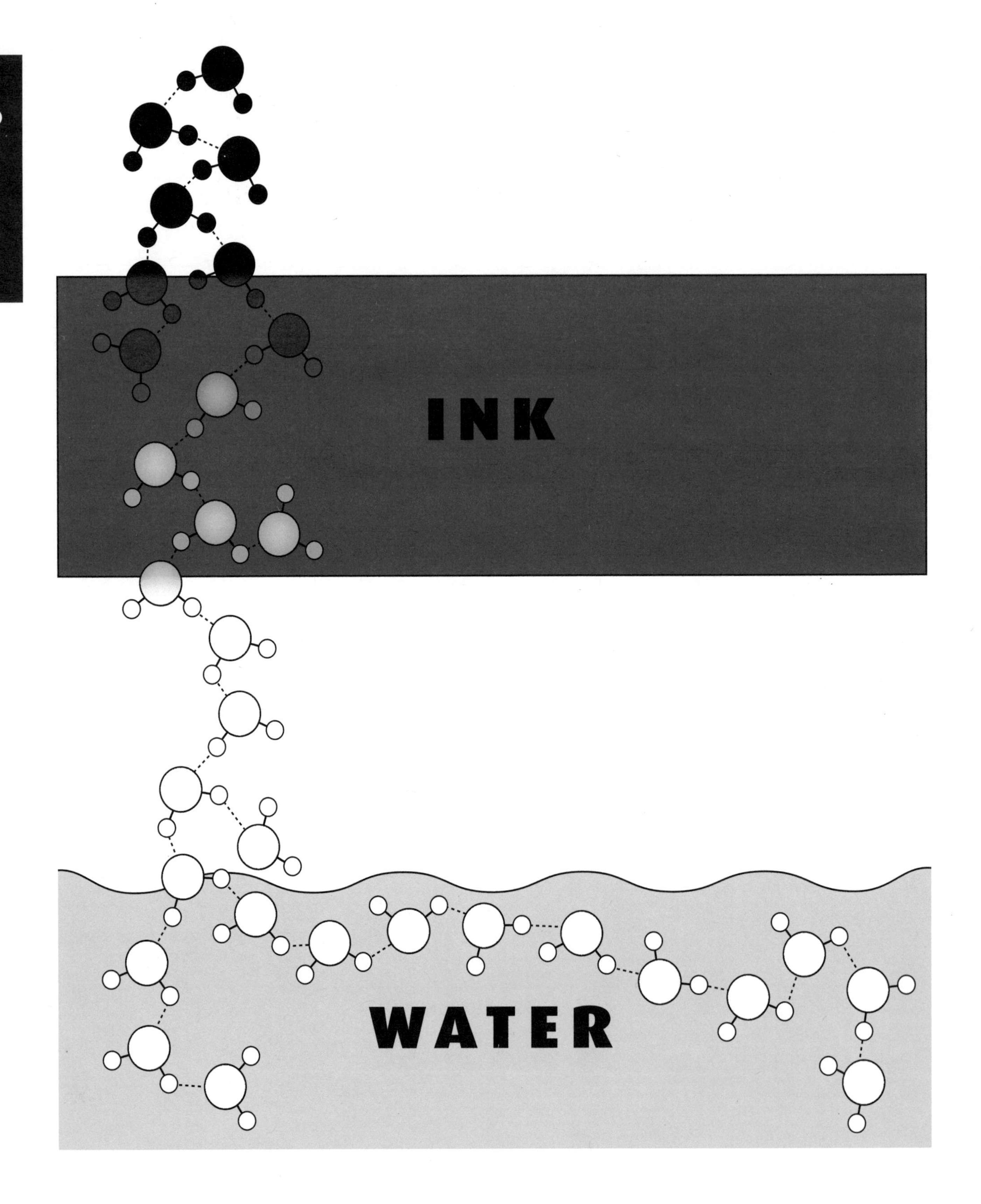

Model A

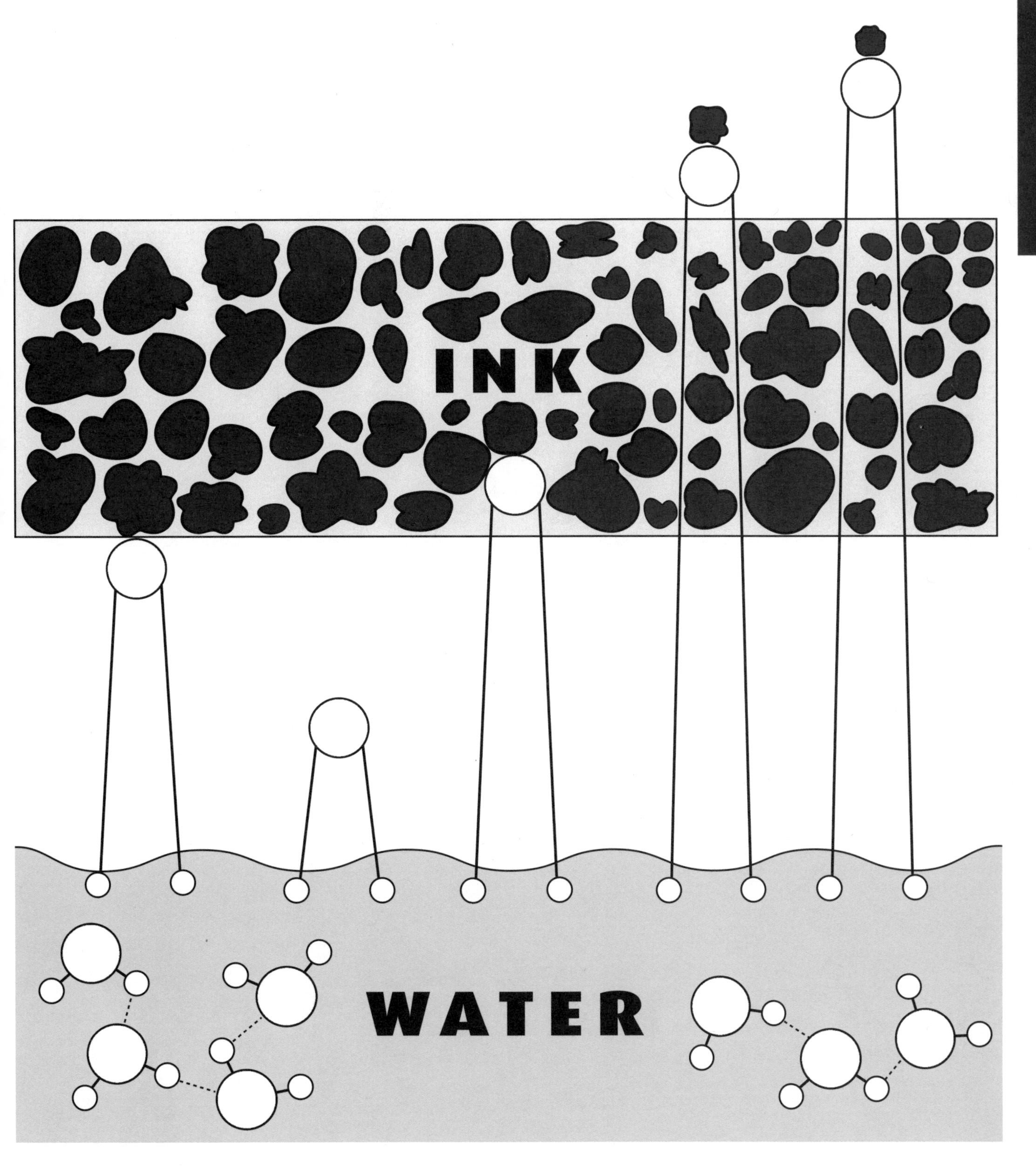

Model B

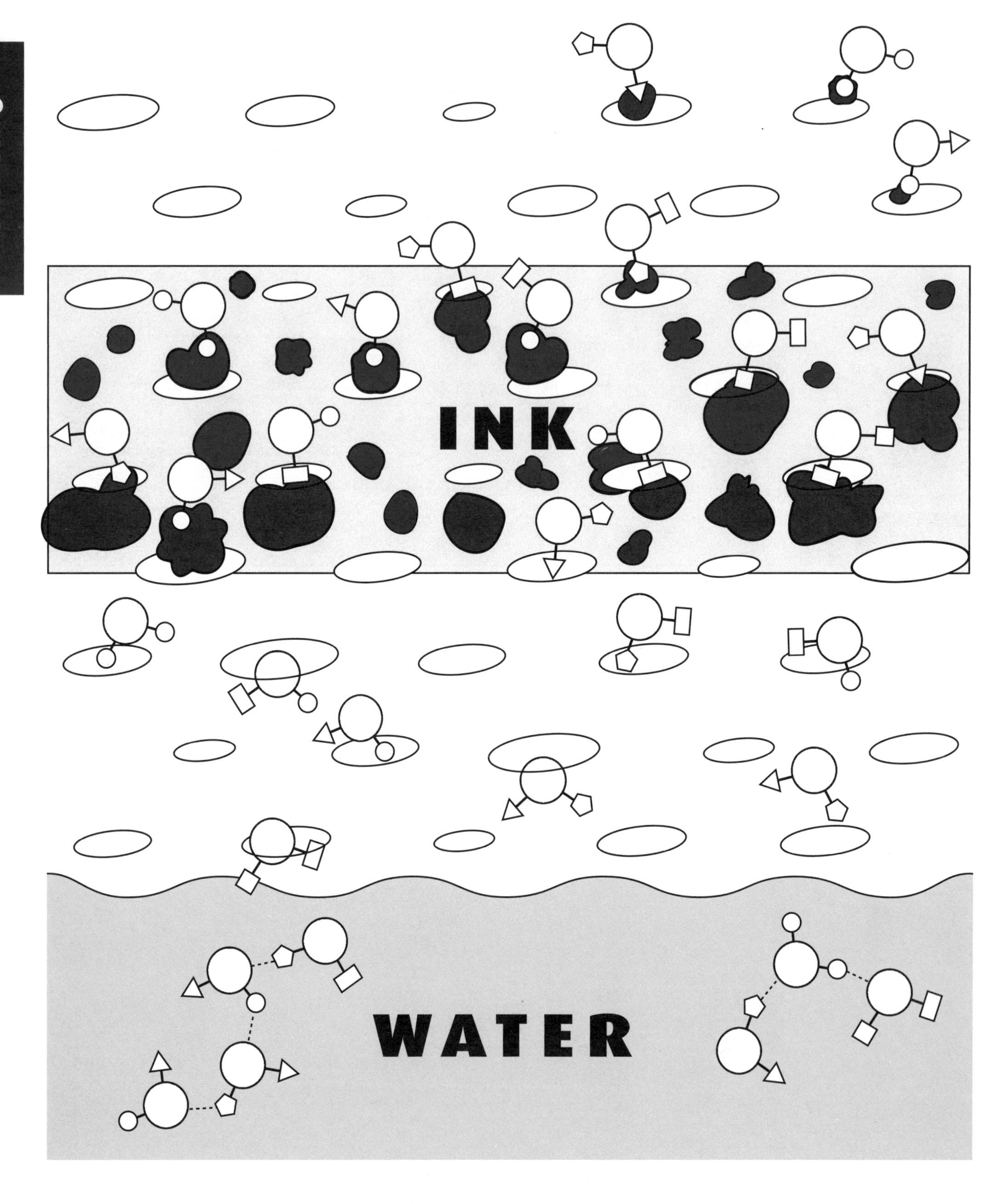

Model C

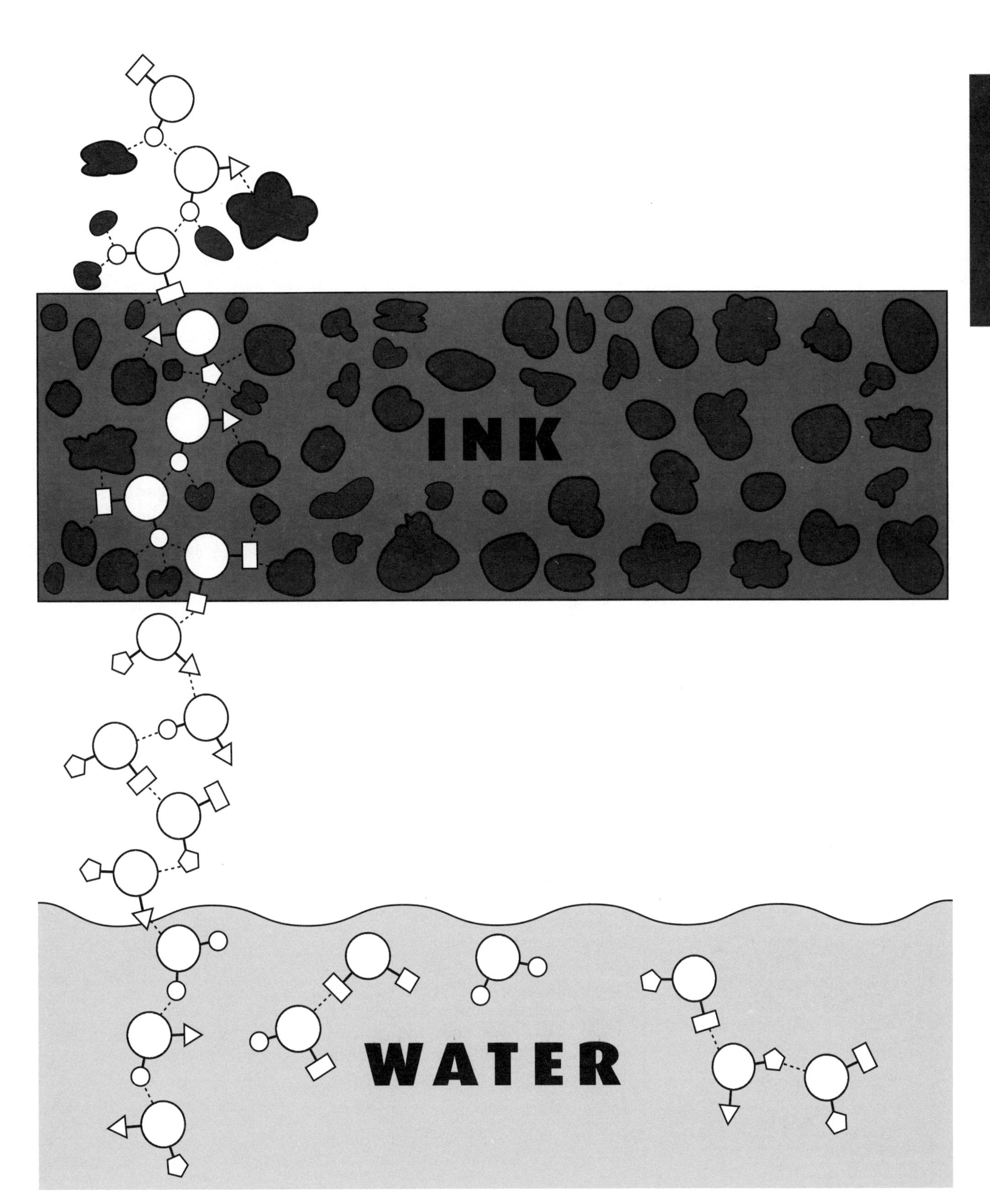

Model D

name

OVERHEAD/STUDENT HANDOUT ACTIVITY 2 • page 1 of 2 •

Four Models

What seems accurate and what seems wrong with these four different models of what happens when ink is separated by water molecules through chromatography?

Things we know about molecules:

- molecules can't change size
- all water molecules look the same
- all water molecules are attracted to each other
- molecules can't stretch
- water molecules can't change colors

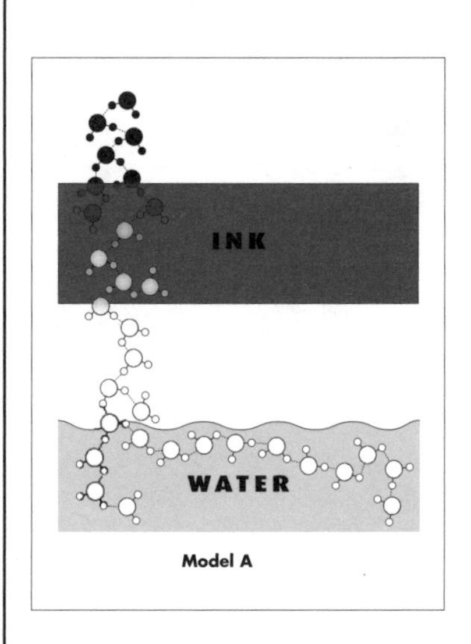 Model A	**Model A** Water molecules are attracted to each other and to fibers in the paper. These attractions cause them to move up the paper. When they go through the ink, they change color.	1. Does it explain how pigments are carried by water molecules? 2. Does it explain why some colors go higher? 3. What parts do you think might be wrong? 4. What parts do you think might be accurate?
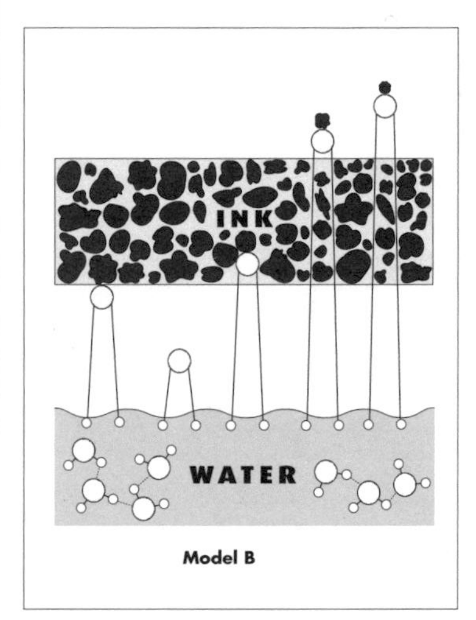 Model B	**Model B** The water molecules stretch as they climb the paper. The lighter weight pigments move higher and the heavier ones less high. The heaviest ones don't move up.	1. Does it explain how pigments are carried by water molecules? 2. Does it explain why some colors go higher? 3. What parts do you think might be wrong? 4. What parts do you think might be accurate?

What seems accurate and what seems wrong with these four different models of what happens when ink is separated by water molecules through chromatography?

Things we know about molecules:

- molecules can't change size
- all water molecules look the same
- all water molecules are attracted to each other
- molecules can't stretch
- water molecules can't change colors

<table>
<tr>
<td>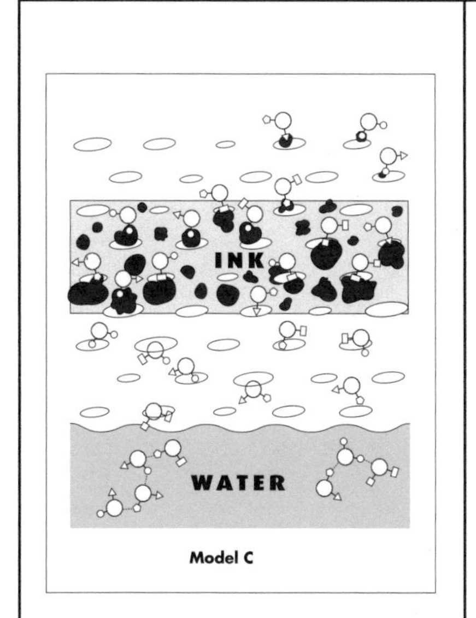

Model C</td>
<td>Model C
The water molecules move up through tiny holes in the paper. The ink pigments are attracted to the water molecules. The bigger pigments can't fit through some holes so they don't go as high.</td>
<td>1. Does it explain how pigments are carried by water molecules?
2. Does it explain why some colors go higher?
3. What parts do you think might be wrong?
4. What parts do you think might be accurate?</td>
</tr>
<tr>
<td>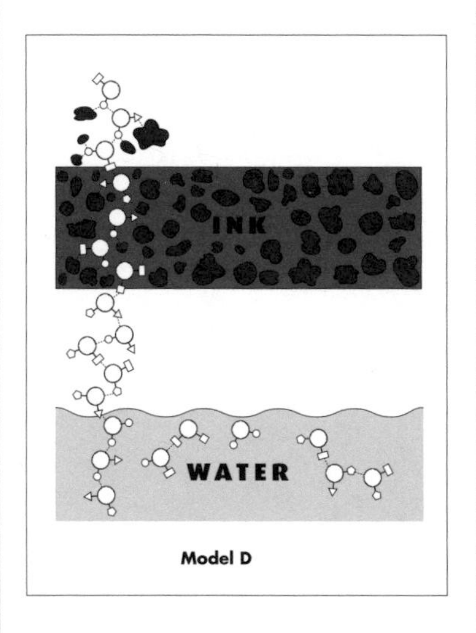

Model D</td>
<td>Model D
The pigment molecules are attracted to the water molecules and to some extent to the paper molecules. The pigments that are more attracted to paper than to water don't go as high. The pigments more attracted to water than paper go higher.</td>
<td>1. Does it explain how pigments are carried by water molecules?
2. Does it explain why some colors go higher?
3. What parts do you think might be wrong?
4. What parts do you think might be accurate?</td>
</tr>
</table>

name

New Microscope Eyes Model of Chromatography

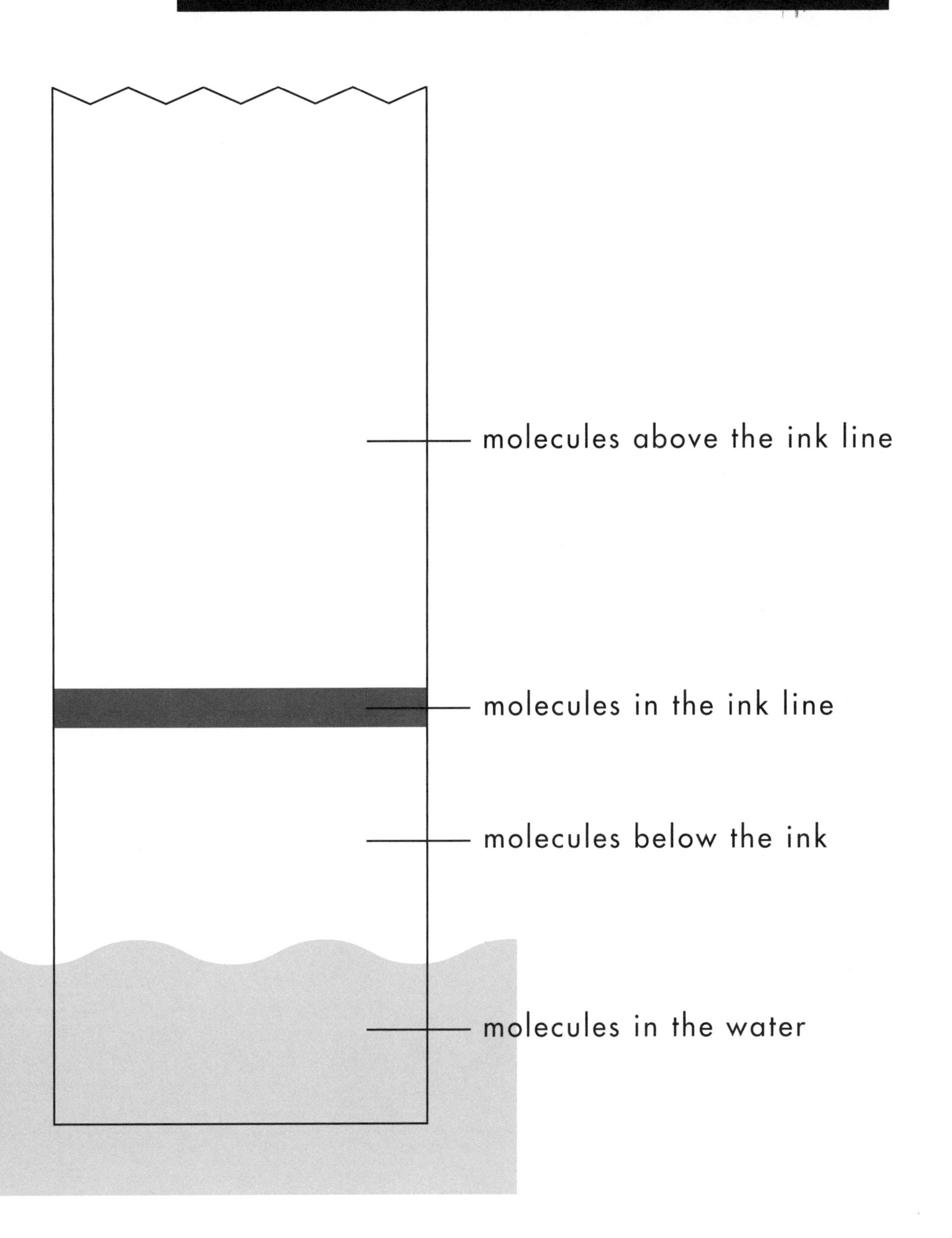

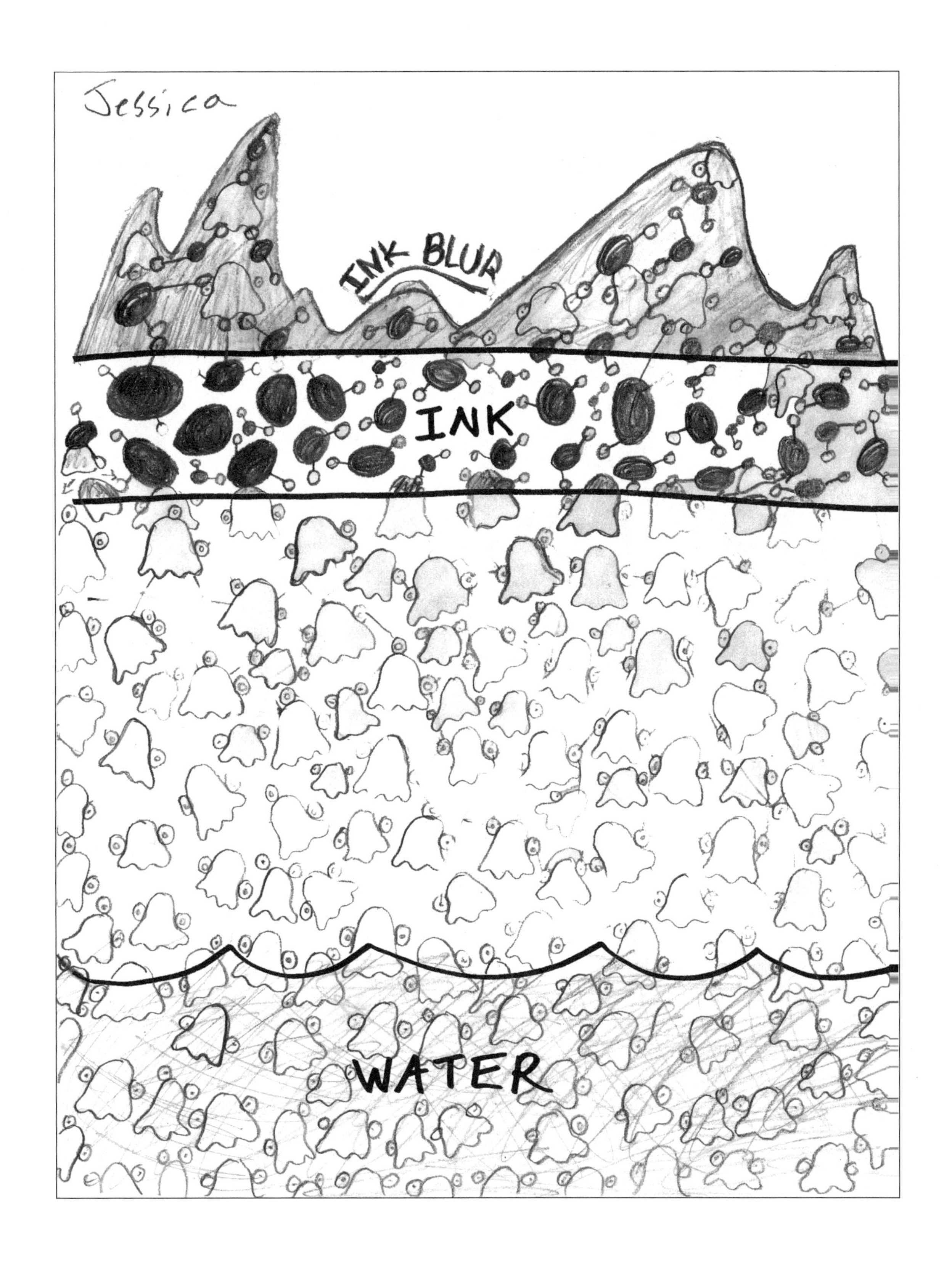
Jessica
INK BLUR
INK
WATER

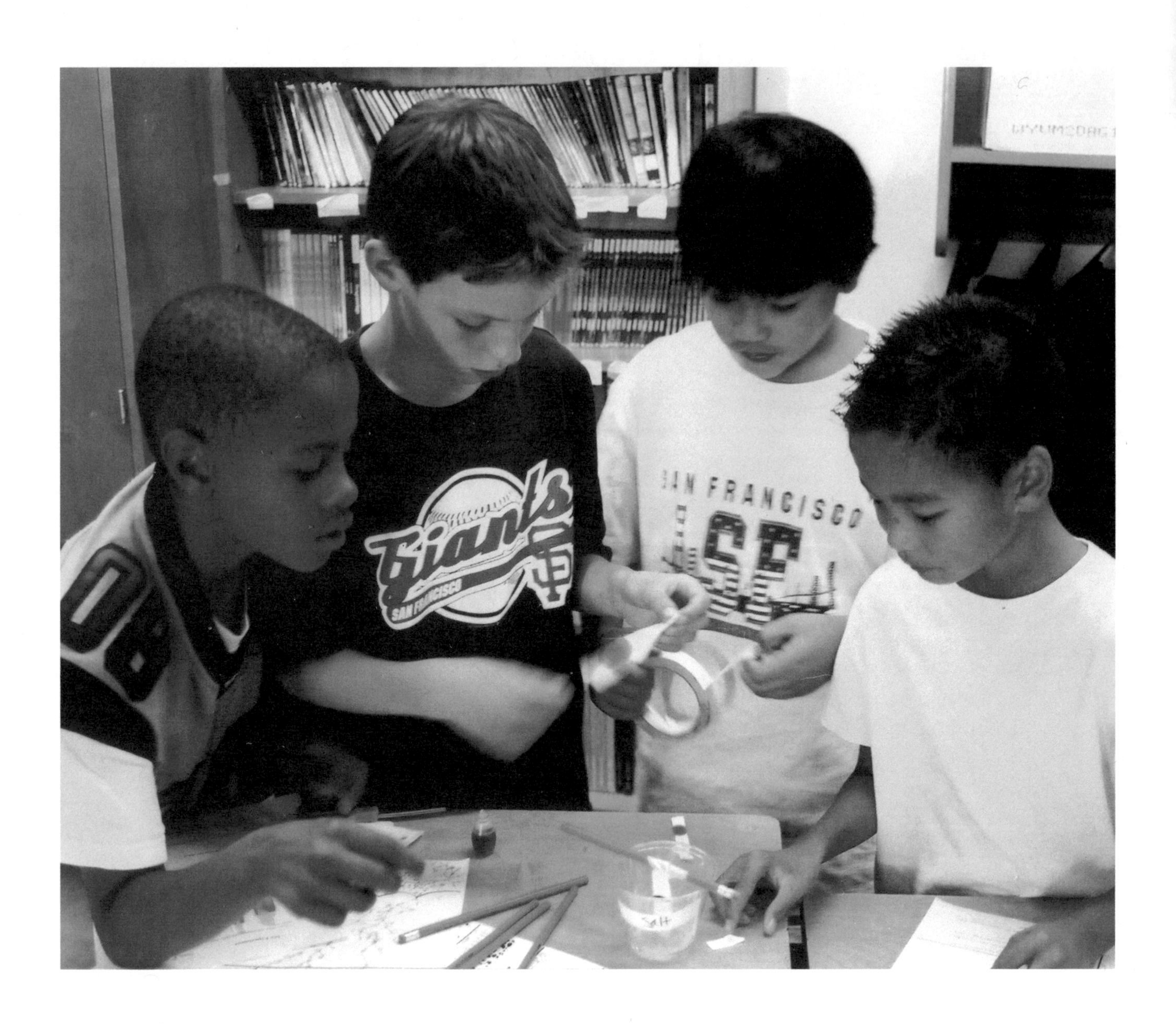
Giants
08
SF

Overview

This activity takes two class sessions. In the first session, students use water and chromatography paper to explore a variety of test substances. Students gain experience using chromatography to separate a variety of substances and wonder, as scientists do, whether another solvent (such as vinegar or alcohol)—or another medium—might work better to separate some of these test substances.

In the second session, students choose one test substance that interests them. They try different combinations of mediums and solvents to design the best system for separating the test substance they've selected. To allow the flexibility needed for students to choose their investigations, three materials stations are set up around the room: Test Substances, Mediums, and Solvents. Students record their test results on a data sheet. They then write an account of their chromatography tests and propose explanations for their results, including a new molecular "microscope eyes" model. This final writing activity can be used as an assessment of students' understanding of the basic ideas of chromatography.

Learning Objective for Activity 3

- Students can act as laboratory scientists, and apply all the concepts and processes learned in the prior class sessions.

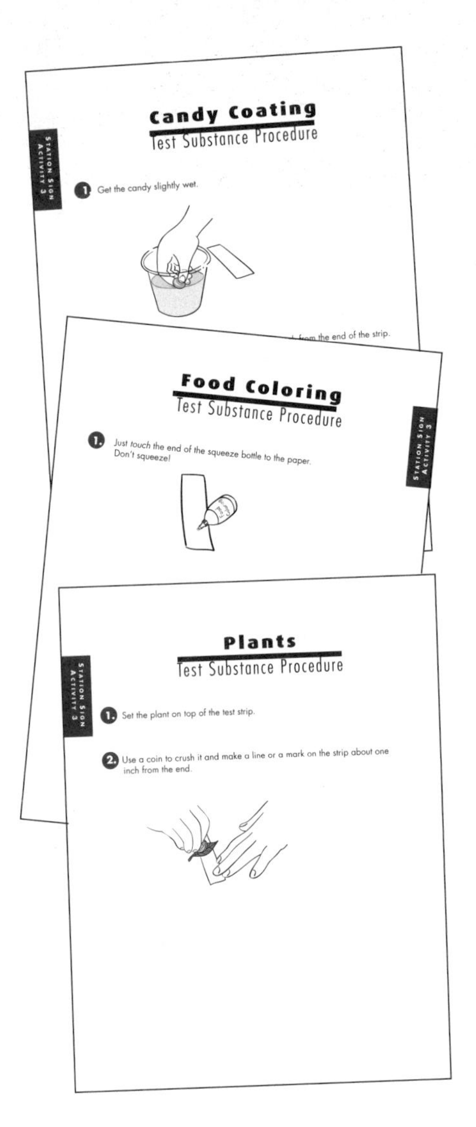

Candy Coating
Test Substance Procedure
Station Sign Activity 3
1 Get the candy slightly wet.
Food Coloring
Test Substance Procedure
Station Sign Activity 3
1 Just touch the end of the squeeze bottle to the paper. Don't squeeze!
Plants
Test Substance Procedure
Station Sign Activity 3
1 Set the plant on top of the test strip.
2 Use a coin to crush it and make a line or a mark on the strip about one inch from the end.

The goal here is to provide a variety of substances for students to test. Don't worry if you can't gather every one of the items listed. For the colored pens, candy coatings, and food colorings, the colors we've suggested usually separate into different colors. However, it can be interesting to include additional colors of pens, candies, and food colorings to allow students to discover this. Please see the "Getting Ready" section for more information on the options for test substances.

Session 1: Choosing a New Test Substance

■ What You Need

For the class:

- ❑ water
- ❑ 1 ruler
- ❑ 1 pair of scissors

For the Test Substances station:

- ❑ 1 copy each of the **Candy Coating** (page 58), **Food Coloring** (page 59), and **Plants** (page 60) **Test Substance Procedure** sheets
- ❑ 2 or more different brands of black water-based felt-tip pens
- ❑ 2 or more different brands of brown water-based felt-tip pens
- ❑ a variety of other colors of water-based felt-tip pens, especially green or purple
- ❑ at least 12 brown and green M&M's®, Reese's Pieces®, or other candies with colored coatings
- ❑ a wide-mouthed cup
- ❑ a 1 oz. squeeze★ bottle of green food coloring
- ❑ a 1 oz. squeeze★ bottle of red food coloring (for mixing with green to make brown)
- ❑ a few different vegetable leaves such as red cabbage, spinach, beet greens, or other dark green leafy vegetables
- ❑ a few freshly picked deep green or red leaves from trees
- ❑ a few freshly picked flower petals that are a deep reddish brown
- ❑ several pennies for crushing plant pigments onto paper strips
- ❑ a few Post-it® Notes or scratch paper to label the vegetables and other plant samples at the station
- ❑ newspaper or plastic trash bags to cover the surface of the station
- ❑ *(optional)* other colors of M&M's or Reese's Pieces
- ❑ *(optional)* additional colors of food coloring

★If squeeze bottles of food coloring are not available, you can pour a very small amount into a cup and add a medicine dropper or a toothpick for students to use to apply the food coloring to the paper strip.

For each group of 4–6 students:

- ❑ about 20–30 4" strips of chromatography paper★★ (or white coffee filters cut into 1" x 4" strips)
- ❑ 1 water trough (plastic wallpaper-paste trough)
- ❑ 2 or 3 pencils (or drinking straws) long enough to set across the water trough

- ❑ a few inches of masking tape
- ❑ a few sheets of white scratch paper
- ❑ pencils

**In Activity 1, chromatography paper was not necessary, since paper towels work well to show the separation of black inks. However, for Activity 3 we recommend using chromatography paper, if possible, because it provides a much better separation, is more effective, and easier to interpret. Some of the subtle colors of candy coatings and plant pigments are difficult to notice without using it. Chromatography paper can be purchased from many scientific supply stores, including Carolina Biological Supply, the distributor of GEMS Kits. For more on obtaining chromatography paper, see page 85. If obtaining chromatography paper is not possible, cut-up coffee filters will work. Coffee filters are a better medium than paper towels.

■ Getting Ready

1. During this session, pairs of students will go back and forth to the Test Substances station where they will apply different substances to their chromatography test strips. To minimize crowding, choose a long counter or long table(s) for the station. Spread newspaper or plastic bags to protect the surface.

2. Gather a variety of substances for the station. To help students succeed and lessen frustration or disappointment, it's best to include many substances that have multiple pigments that separate and are easy to see, and fewer substances that do not. Here are some tips:

   a. **Different colored water-based pens.** Browns, blacks, and greens usually have the most different-colored pigments. Different brands of these colors are interesting to test, along with other colors.

   b. **Candies with colored coatings.** We suggest you provide only brown and green samples because they contain multiple pigments. You can have other colors on hand if students request them. Sort out the green and brown colored candies, and if you are using a brand other than M&M's, pre-test them to make sure the colors separate.

   c. **Food coloring.** Both green and brown food coloring work well because each separates into more than one color. You can create brown food coloring by mixing two tablespoons of red food coloring with one tablespoon green.

To prevent rampant munching, many teachers decide not to set the candies out at the station, but rather keep them and hand them out, one green and one brown per group, as students request them.

d. **Plant pigments.** These pigments are quite subtle, so choose reddish brown or other dark colored leaves, vegetables, or flower petals. If you include a green leaf from a tree that you know will change color in fall, the chromatograms may reveal hidden red, orange, or yellow pigments. Have about half a dozen pennies ready to set out near the plant samples. Students will use the coins to crush the plant pigments onto the test strips. Make a small sign to place near each plant sample as a label (for example, "red cabbage").

3. Fill the wide-mouthed cup about half full with water. This will be used to get the candy slightly wet so students can spread the candy's color onto the strip of chromatography paper.

4. Set out the materials at the Test Substances station.

5. Make one copy each of the **Candy Coating** (page 58), **Food Coloring** (page 59), and **Plants** (page 60) **Test Substance Procedure** sheets. Place each sign near the appropriate substances at the station. You may need only one copy of each sign, but, for example, if you have more than a few plant pigments, it's wise to have several procedure signs for plants. You won't need a procedure sign for the colored pens because students learned that procedure in Activity 1.

6. Decide how desks will be arranged so that students can share materials in groups of from four to six students.

7. Prepare the following materials and have them handy for distribution to the groups after you have introduced the activity.

   a. For each group of four to six students, fill a water trough with no more than $^1/_2$ inch of water.

   b. Cut enough 4" strips from the chromatography paper that each *pair* of students will have about 10 strips. (If you are using coffee filters, cut 1" x 4" strips as in Activity 1.)

   c. Have masking tape, pencils, and sheets of scratch paper ready.

■ Introduce the Activity

1. If necessary, briefly review what chromatography is used for [separating mixtures] and the parts of a chromatography system [medium, solvent, and test substance].

2. Explain to students that they'll use water in a trough again, as a solvent. The medium will be strips of chromatography paper (or coffee

filters). If your students did not use chromatography paper in Activity 1, explain that it is especially designed to show clear separation of substances.

3. Ask what substances they could test other than the inks from black felt-tip pens. [Mention the following ideas if they do not: Different colored pen inks, the coloring in candy coating, food coloring, and the color in plant leaves or flowers.]

4. Show students the items at the Test Substances station. **Emphasize that they don't need to test all the substances.** They should test a few that interest them, **with the goal of finding one test substance to investigate further** in the next session.

5. Tell them everyone at their table group will share a water trough, strips of chromatography paper, pencils, tape, and scratch paper. But, when they do their chromatography tests, they will be working with a partner. Assign partners now to avoid distraction as you are explaining the procedure.

6. Tell students that partners will take one strip of chromatography paper and go to the Test Substances station. They will choose one test substance, read the procedure for it, and apply a bit of it to their paper strip. They should then take the strip to their desk to do the test, being sure to leave the test substances at the station.

7. Let students know that each time they finish a test, they will observe the results closely, tape their chromatogram to a sheet of scratch paper, and write near it which test substance they used. (Both partners' names should be on the scratch paper.) Then they can get another strip of chromatography paper, go to the station again with their partner, and choose a different test substance.

■ Demonstrate How to Apply Test Substances

1. Demonstrate how to apply different test substances to the paper strip (medium):

   a. Say that if they choose ink as a test substance, they'll apply it to the paper as before by drawing a line about one inch from the end of the strip.

   b. When testing the candy coatings, they should slightly wet the candy with water, then smear the coloring onto the paper strip in a line about one inch from the end of the strip.

c. When testing food coloring, they should just touch the end of the squeeze bottle to the paper *without squeezing* and try to make a **tiny** mark about one inch from the end of the strip. Be sure students know they should **not** use a whole drop on the paper—it's too much, and their chromatography test will not work well.

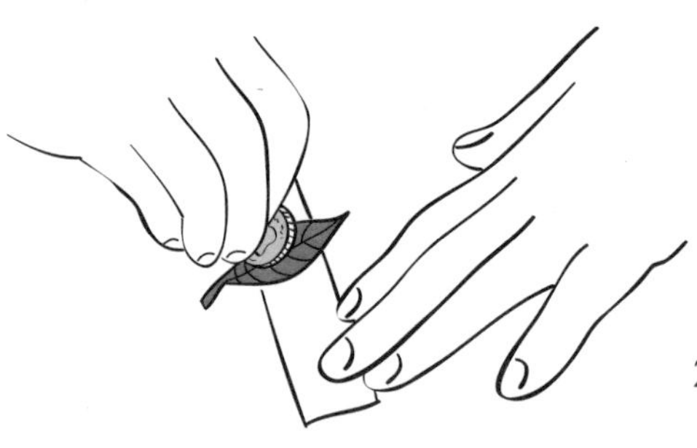

d. When testing plant leaves or flower petals, they should set the test substance on top of the paper strip, then use a coin to crush it, creating a line or a mark on the paper about one inch from the end of the strip.

2. Mention that there are three procedure signs at the station to remind them of what you've just demonstrated.

If you decided to keep the candies with you rather than setting them out at the station, let students know that now.

3. Describe to students the calm, cooperative behavior that you expect of them. One way to review the correct procedures is to demonstrate the "wrong way," and have them tell you what the right procedure would be. For example:

- Run from desk to station.
- Take lots of test strips to the station at once.
- Grab samples from the station and take them to your desk.
- Ignore your partner.

Students Test Substances

1. Distribute troughs, test strips, and other materials to the groups. Choose the first few pairs of students to go to the Test Substances station, then allow everyone to go.

2. Circulate and make sure students observe results, tape their completed chromatograms to the scratch paper, label each one with the test substance they used, and write their names on the scratch paper.

3. When all groups have had a chance to test at least three substances, have them sit down. Then collect the materials from table groups. Also collect their papers with chromatograms taped on them. Have someone straighten up and re-organize the Test Substances station, and leave it set up for the next session.

4. Tell the class they will get to share their results and do further tests in the next class session.

Session 2: Designing Chromatography Systems

■ What You Need

For the Test Substances station:

- ❑ the station materials from the previous session
- ❑ *(optional)* a few different colored permanent pens

For the Mediums station:

- ❑ chromatography paper
- ❑ white coffee filters
- ❑ different brands of white paper towel
- ❑ white cotton cloth
- ❑ white construction paper
- ❑ toilet paper or tissues
- ❑ white notebook paper
- ❑ a few pairs of scissors

It is not necessary to provide all the different types of paper and cloth listed here. Just provide as much variety as is practical for you. Since chromatography paper and coffee filters are likely to be in demand, provide more of those test strips. See the "Getting Ready" section for more information.

For the Solvents station:

- ❑ about 6 water troughs from previous sessions
- ❑ about 24 wide-mouthed cups
- ❑ a permanent pen to label cups
- ❑ 1 tablespoon of baking soda
- ❑ 1 tablespoon of salt
- ❑ 1 quart or liter bottle of white vinegar
- ❑ 1 quart or liter bottle of rubbing alcohol
- ❑ water
- ❑ *(optional)* clear plastic wrap to cover the rubbing alcohol cups

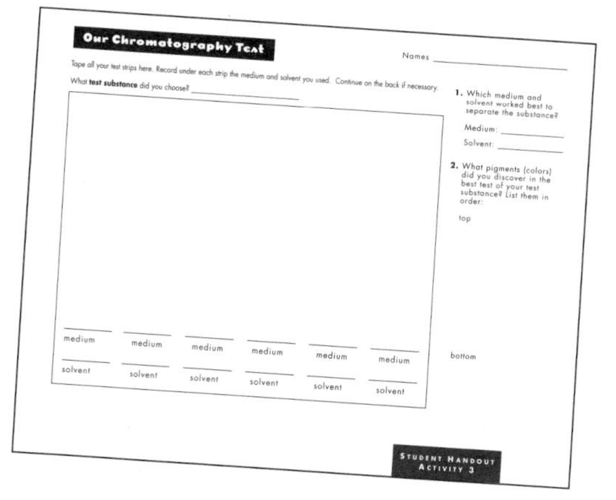

Our Chromatography Test

Names ____________

Tape all your test strips here. Record under each strip the medium and solvent you used. Continue on the back if necessary.

What **test substance** did you choose? ____________

1. Which medium and solvent worked best to separate the substance?
Medium: ________
Solvent: ________

2. What pigments (colors) did you discover in the best test of your test substance? List them in order:
top

bottom

medium	medium	medium	medium	medium	medium
solvent	solvent	solvent	solvent	solvent	solvent

STUDENT HANDOUT ACTIVITY 3

For each group of 4–6 students:

- ❑ 2 or 3 pencils (or drinking straws) long enough to set across the water trough
- ❑ a few inches of masking tape
- ❑ pencils

For each pair of students:

- ❑ 1 copy of the **Our Chromatography Test** sheet (page 61)
- ❑ their scratch paper with chromatograms from the previous session

For each student:

- ❑ 1 copy of the **Chromatography Test Report** sheet (page 62)

■ Getting Ready

1. Make one copy for each *pair* of students of the **Our Chromatography Test** sheet (page 61). Make one copy for *each student* of the **Chromatography Test Report** sheet (page 62).

2. Have handy the masking tape and several pencils for each group of four to six students. Also have available the papers with chromatograms from the previous session.

3. As described below, set up the Test Substances, Mediums, and Solvents stations on separate tables in areas where they will be accessible during the activity.

a. For the Test Substances station, use the same set up as in the previous session. If you've decided to use them, set out some permanent pens. The permanent ink will sometimes separate with alcohol or vinegar, but the chromatogram will usually be shades of the color.

b. For the Mediums station, cut about a dozen of each of the medium materials into 1" x 4" strips. You may want to cut extra strips of the chromatography paper, since they are likely to be heavily used. Set out the test strips in distinct piles at the station. Since it is difficult to anticipate how many strips will be needed, also set out some of the uncut medium materials and scissors at the station so students can cut strips themselves if necessary.

c. For the Solvents station, set up the following materials:

- Fill the water troughs with no more than $^1/_2$ inch of water.
- Use the permanent pen to label six cups for each of the other four solvents: baking soda solution, salt water, vinegar, and rubbing alcohol.
- Mix one tablespoon of baking soda into a cup of water, and stir till the baking soda has dissolved. Pour $^1/_4$ inch into each of the six baking soda solution cups.
- Stir one tablespoon of salt into a cup of water. Pour $^1/_4$ inch into each of the six salt water cups.
- Pour $^1/_4$ inch of vinegar into the bottom of the corresponding labeled cups. Do the same with the rubbing alcohol. *Note: Rubbing alcohol evaporates quickly. Plan to pour it into the cups right before the activity, or pour it earlier but cover the cups with clear plastic wrap.*

■ Discuss Test Substance Results

1. Distribute the papers with chromatograms from the previous session to pairs of students. Ask a few students to share their results. As they share, ask what they found interesting and what kind of separation they observed, if any.

2. Ask, "What substances didn't separate at all or were difficult to see?" Ask what they could do differently to get better separation. [Try different combinations of mediums and solvents to find the best system for separating a test substance.]

■ Introduce the Activity

1. Explain that, based on their test substance results, each pair of students will choose **one test substance** from the Test Substances station that is most interesting to them. They will get to design their own chromatography system with the best medium and solvent to get the best separation they can.

Students may assume that chromatography paper is always the best medium to choose. Tell the students that chromatography paper does help show different colors that are hard to see on other mediums. However, it may or may not be the best medium to use, depending on their test substance.

2. Ask for ideas on what mediums they might use other than chromatography paper. [Mention the following if they do not: chromatography paper, different brands of paper towels, coffee filters, toilet paper, tissue paper, fabric, construction paper, notebook paper.] Show them the test strips available at the Mediums station and let them know that there are extra materials and scissors available for cutting their own strips if the supply has run low.

3. Ask what solvents they might use other than water. [Mention the following ideas if they do not: baking soda solution, salt water, vinegar, rubbing alcohol. Say that other solvents like nail polish remover or ditto fluid might work, but they are not provided because they give off fumes.] Point out what is available at the Solvents station. If you've covered the cups of rubbing alcohol, explain to students that this is to prevent evaporation.

4. If you've added permanent pens to the Test Substances station, let students know about them.

■ Designing Chromatography Systems

1. Tell students that they will start by looking over their chromatograms from the previous session, then decide with their partner on **one** test substance that both partners think would be interesting to test further.

2. Hold up an **Our Chromatography Test** sheet that pairs will share. Demonstrate the procedure briefly, using a student as your "partner."

   a. Say you and your partner have agreed that red cabbage (or whatever you wish) will be your test substance. Show where you would write "red cabbage" on the **Our Chromatography Test** sheet.

   b. Say that next you and your partner will predict which medium and solvent might work to separate the pigments in red cabbage. Then the two of you can go to select a test strip from the Mediums station. Take the strip to the Test Substances station and apply some red cabbage with a coin.

   c. Next say that you and your partner will go to the Solvents station to get a cup (or trough) of the solvent you've chosen. Then you'll take it to your table to do the test.

   d. Tell students that after they do each test, they will tape the chromatogram to the **Our Chromatography Test** sheet, *including* those that don't show separation of the test substance well.

   e. Show where they will note which solvent and medium they used under each test strip. They may need to go back to the Mediums and Solvents stations to get more materials to do different tests to find the best system for separating their test substance. They are to return their solvent to the station before they try another solvent.

   f. Tell students that when they have found the best separation of the test substance, they'll record the medium and solvent used and respond to the other questions on their sheets.

3. Hand out the **Our Chromatography Test** sheets to each pair of students. Give some masking tape and a few pencils to each table group.

4. Once student pairs have chosen and recorded their test substance, allow them to go to stations as needed.

5. Circulate to ensure that partners are working together, following the test procedures, and recording their results after each test.

Discussing and Reporting Results

1. When students have finished their tests and completed the questions on their sheets, reconvene for a class discussion.

2. Take a few minutes to have a few students share what test substance they chose, what mediums and solvents they tried, and why they think they got some of the results they did. What were some surprises? Frustrations? Did they need to change their focus midway?

This assignment will allow students to reflect more fully on their test results, and also assesses their ability to use a molecular model to explain chromatography test results.

3. Congratulate them on conducting themselves just like scientists.

4. Say that each student will now get a **Chromatography Test Report** sheet on which to record the details of what she found out. Go over the questions on the sheet, and emphasize that their explanations need not be "correct," but do need to be based on evidence.

5. Point out the place where they will draw a "microscope eyes" model that shows what they think happened to the molecules of the test substance, medium, and solvent during their most successful test. Let students know their model should try to respond to questions such as:

    - Were there many or few colors visible?
    - Did certain colors travel higher or lower? Why?
    - Did certain colors not travel at all? Why?

6. Pass out a **Chromatography Test Report** sheet to each student and have them begin.

7. Have a few early finishers clean up and organize the station materials.

8. When students have finished, collect the **Chromatography Test Report** sheets. As appropriate, after you've had a chance to review their work, discuss their responses, especially to clarify any persistent misunderstandings or other issues that arise. Encourage students to raise ongoing questions they may have.

Candy Coating

Test Substance Procedure

1. Get the candy slightly wet.

2. Smear a line of the candy coloring about one inch from the end of the strip.

Food Coloring

Test Substance Procedure

1. Just *touch* the end of the squeeze bottle to the paper.
Don't squeeze!

2. Make a **tiny** mark about one inch from the end of the strip.
A whole drop is too much.

Plants

Test Substance Procedure

1. Set the plant on top of the test strip.

2. Use a coin to crush it and make a line or a mark on the strip about one inch from the end.

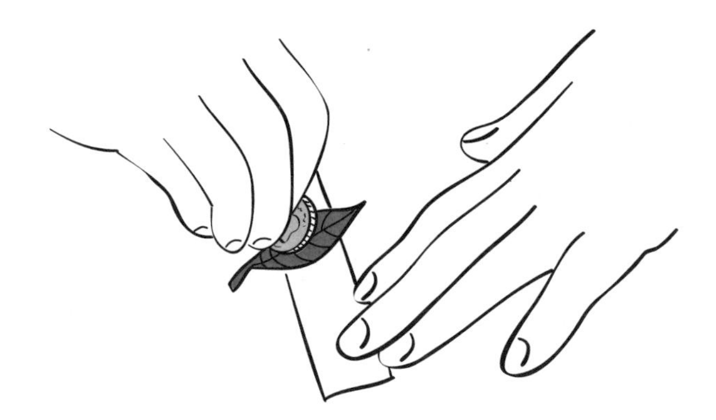

Our Chromatography Test

Names ____________________

Tape all your test strips here. Record under each strip the medium and solvent you used. Continue on the back if necessary.

What **test substance** did you choose? ____________________

medium	medium	medium	medium	medium	medium
solvent	solvent	solvent	solvent	solvent	solvent

1. Which medium and solvent worked best to separate the substance?

Medium: ____________

Solvent: ____________

2. What pigments (colors) did you discover in the best test of your test substance? List them in order:

top

bottom

STUDENT HANDOUT
ACTIVITY 3

Chromatography Test Report

Name ____________________

1. Test substance: ________________________

Which medium and solvent worked best to separate it?

Medium: ______________________ Solvent: ________________________

2. Pretend you have "microscope eyes," and draw a model in the box to show what you think happened to the molecules of the test substance, medium, and solvent listed above. Write a description to go with your drawing.

3. Why do you think you got the results you did? Remember to list evidence to support your ideas. For example, the evidence could include the results you got with other mediums or solvents, as well as which did or did not separate the test substance.

(Continue on the back of the paper if needed.)

1. Write a Mystery

Assign students to write their own versions of the mystery/crime in Activity 1. If you are teaching this unit to multiple classes, one class could write a mystery for the next class to solve.

2. Tell the Crime Story

Ask your students to focus on who they think may have committed the crime in Activity 1, then write their own version of the story from a distinct point of view. This could be from the perspective of a detective or reporter investigating the case, one of the suspects, or an impartial bystander. Tell students that although they can get creative with the story and details, it must also be based on the evidence of the chromatograms.

3. Create a Mystery Scenario

After Activity 3, when students have experienced more test substances, solvents, and mediums, have student groups create other mystery scenarios with different test substances such as candy coatings, colored pens, or plant leaves as the key to solving the mystery. Have them set up materials and have other teams of students try to solve the mystery.

4. Literature Connections

Reading mystery stories is a great literacy accompaniment for this unit. See the "Literature Connections" section for ideas, or work with the school librarian.

5. Chromatography Art

For an interesting and artistic extension, some teachers have students draw a water-color pen design on a coffee filter, then fold the filter into a cone, and dip the tip in water. Students can watch the colors spread and the filters can be opened and hung in the window for decoration. These can also be made into flowers, with pipe cleaner stems. Students can also do drawings with permanent pens on coffee filters or paper towels, then add different brands of water-based black felt-tip pens that they know will separate into colors with water. The color patterns of some black water-based pens resemble flames or beautiful sunsets. When the bottom portion of the paper is set in water, the water-based pen ink adds its range of colors to the drawing.

6. Continue with Related GEMS Units

Present these *Crime Lab Chemistry* activities in the extended context of other GEMS units, such as *Fingerprinting* and *Mystery Festival*. Dr. Steve Rakow, a former president of the National Science Teachers Association (NSTA), an outstanding science educator, and close friend of GEMS, used *Crime Lab Chemistry* and *Fingerprinting* as a "forensic science academy." Students would go through this academy and receive a certificate to be forensic scientists. Then—as forensic scientists—they would take part in the GEMS *Mystery Festival* unit! Steve died tragically in 2001 as a result of a bicycle accident, but his spirit of inquiry lives on!

Background for the Teacher

The following is intended to provide you, the teacher, with information that should be helpful in responding to student questions. It is not meant to be read out loud or duplicated for students. You and your students may want to do more research on chromatography, inks, or pigments. See also the "Resources and Literature Connections" section starting on page 85.

More about Chromatography and Pigments

Where does the term chromatography come from?
The word derives from Greek for "chroma" (color) and "graph(y)" (writing). It was first used in the context of chemistry in 1906 by a Russian botanist, Mikhail Tswett (also spelled Tsvett or Tsvet). He filled a column with fine calcium carbonate to separate plant pigments. Although chromatography is now used to separate many substances in many other ways that don't necessarily relate to color, the scientific term comes from its original use with plant pigments.

Why does the water travel up the paper?
It is due to capillary action.

What is capillary action?
Capillary action refers to the ability of water to be drawn upward, as for example in the stems of plants. It is a process involving both cohesion and adhesion. Water molecules tend to be attracted to each other. A water molecule is made of two positive hydrogen atoms on one end and one highly negative oxygen atom at the other. This structure enables water molecules to stick together, with the negative ends of one molecule next to the positive ends of another. When molecules are attracted to "like" molecules (such as one water molecule to another water molecule), it is called ***cohesion***. Cohesion between water molecules is responsible for surface tension. At the same time, water molecules are not only attracted to each other, they are also attracted to many other types of molecules. When molecules are attracted to unlike molecules, the process is called ***adhesion***. Adhesion is why water is attracted to things that aren't water (like us, when we get wet). Because water molecules are also attracted to paper towel fibers (adhesion) the water "climbs" the paper. But because the water molecules are still attracted to each other (cohesion) other water is pulled up. Due to both cohesion and adhesion the water keeps moving up the paper until the pull of gravity is too much for it to overcome. This upward pull is called capillary action.

What is chromatography?

Chromatography is an important technique in chemistry for separating out the components of mixtures. There are numerous instances during chemistry research and experimentation when such separation is useful or needed and chromatography is one of the most effective and frequently used methods. There are many kinds of chromatography.

Chromatography is based on the idea that each type of molecule of a specific substance has its own unique ways of sticking to and detaching from different surfaces and/or materials. A mixture puts different substances together, and each substance has its own unique molecular characteristics. The object of chromatography is to separate the substances.

In chromatography, either a gas or a liquid is selected as a solvent to carry the test substance/mixture along a column. The column is coated with some substance the molecules in the test substance will be attracted to—this is the medium. As the mixture moves along, each type of molecule in the mixture attaches to the column's surface, depending on how strongly each type is attracted to the medium or to the solvent. Those parts of the mixture more attracted to the solvent than to the medium will move along the column more quickly or further than those that are more attracted to the medium than the solvent. The components of the test substance end up separated out according to how well or how poorly they adhere to the medium in the column. The trick is to find the right medium and solvent to separate a specific mixture. How quickly or how far the parts of the mixture move along the column is also affected by other attributes of the molecules such as size, weight, shape, charge, interaction with the solvent, how easily they turn to gas, or the temperature at which they are adsorbed.

Adsorption *is defined as adhesion in a thin layer of molecules (gas or liquid) to a surface. (This is distinguished from absorption.)*

Any substance that can be dissolved in a liquid or evaporated into gas can be separated through chromatography. You just have to find the right medium and solvent. There are a huge number of possible mediums and solvents. When attempting to separate out the components of a mixture, a chemist may need to try many different combinations of possible substances as medium and solvent before finding the combination that separates out the components. The solvent in chromatography is sometimes also called the mobile phase (because it moves through the column) and the medium is sometimes called the stationary phase (because it stays in place).

As your students learn in this unit, the technique of chromatography is often thought of as a system, involving the test substance, the medium, and the solvent. Together, they are a set of connected parts that interact with and affect each other.

What other types of chromatography are there?
There are *many* different kinds of chromatography. Within each, there are a range of strategies and special techniques to enable precision and effectiveness. Here are some of the main forms chromatography currently takes:

Paper Chromatography. This is the type of chromatography used by the students in this unit. A small amount of test substance is placed toward one end of a piece of paper. It is usually set up inside an enclosed container. The end of the paper is placed in a solvent, which moves up the paper through capillary action, separating out components of the test substance. In the paper chromatography technique used in the first activity of this guide, the medium was the paper and the solvent was the water. The pigments that end up closer to the top of the strip of paper are pigments that have less of a tendency to attach to the molecules in the paper and more of a tendency to attach to water molecules. This is the primary cause of separations that happen during paper chromatography. However, the pigments near the top might also be pigments with molecules that weigh less so they rise higher, or pigments with smaller molecules so they fit more easily through the pores in the paper. In paper chromatography, the test substance must interact with the paper, but it cannot chemically react with it to form bonds. If a test substance chemically reacts with the paper it cannot be separated with this approach. Paper chromatography often takes hours to complete. Sometimes the separated components can be seen, and sometimes not. An ultraviolet (UV) lamp is often used to see organic compounds that cannot be seen otherwise. In two-way paper chromatography, a substance is separated first with one solvent, the paper is rotated 90 degrees, then it is separated with a different solvent.

Thin-Layer Chromatography (TLC). This type of chromatography is similar to paper chromatography. Some type of granular material (usually silica gel or alumina) is coated on a glass plate. The test substance is placed near the bottom of the plate. The end of the plate is placed in a shallow pool of a selected solvent. As the solvent travels up the plate by capillary action, it passes the test substance, carrying along with it those parts of the test substance that are soluble in that solvent. How far a particular substance is carried up the plate depends primarily on the balance between its attraction to the solvent (solubility) and its attraction to the medium (adhesion). If the test substance is colored, it is easier to see the results. If it is not colored, ultraviolet light is used. The plate is designed so it fluoresces, and the test substance shows up under ultraviolet light because it does not fluoresce. Thin-layer chromatography is very useful because it can be done with a very small amount of test substance, and is quick and inexpensive. Because different mediums can be used, wider separations can be achieved over less distance than in paper chromatography.

A brilliant woman scientist named Erika Cremer was the first to develop gas chromatography, in Austria in the 1940s. For an excellent summary of her scientific accomplishments in the face of discrimination against women in science, we recommend Women in Chemistry, *published by the Chemical Heritage Foundation (see full listing in the "Resources and Literature Connections" section, page 86).*

Gas Chromatography (GC). In gas chromatography, the test substance is heated until it vaporizes. It is then injected into a chromatography column. The entire column is usually kept in a thermostat-controlled oven to make sure the components don't cool and turn to liquid. The test substance is moved through the column by an inert gas (a gas that will not react with other chemicals) such as helium or nitrogen. The components of the test substance separate out as they attach to the medium. There are two types of columns used in gas chromatography. One is a packed column, where the medium is a packed fine inert solid (such as diatomaceous earth) coated with a liquid, or is a solid material by itself to which the molecules of the test substance/mixture are attracted. The second type of column is a capillary column. In this case, the medium is a coating on the inside of a narrow tube. There is a detector at the other end of the column that measures each component that separates out from the mixture and records the time it arrives at the end.

Liquid Chromatography (LC). Liquid chromatography is used to separate components of liquids. The liquid solvent is sent though a column with a medium (often packed particulate material). The molecules of the test substance that attach most strongly to the medium travel through the column more slowly, while those that attach less strongly travel through more quickly. The medium in liquid chromatography can be either solid or liquid.

High Performance Liquid Chromatography (HPLC). In this type of chromatography, small solid particles are used as the medium. The test substance is moved through the column using a liquid at high pressure. This causes the test substance to have very little time in contact with the medium, which causes more separation between substances.

The following are not technically forms of chromatography, but are related procedures:

Electrophoresis. Electrophoresis is a technique very similar to chromatography that is also used to separate different kinds of molecules. The mixture (test substance) is placed at one end of a gel (the medium), and an electric current is applied causing the solvent or mobile phase (an ionic buffer or charged solution) to move toward the positive end of the gel. The solvent carries the negatively charged molecules with it. The more negatively charged and small a molecule is, the faster it moves through the gel. After they've been separated, the molecules are stained, so they can be seen. Among other things, electrophoresis is used to separate DNA fragments.

Spectrophotometry. Spectrophotometry operates on the principle that different substances absorb and reflect different wavelengths of the electromagnetic spectrum. A beam of light is directed at a sample, and an instrument measures how much light passes through and is not absorbed by the substance. The substance can be identified by the rate of absorption of light.

What are some of the other uses for chromatography?
Chromatography can be used to separate out any substance that can be dissolved or vaporized, with the appropriate medium and solvent. It is used for separating, identifying, and purifying chemicals, so has many uses in the world of science. Here are a few:

- Quality control testing to make sure a product is pure.
- Testing for pollutants in air and water.
- Testing for pesticides in the environment.
- Testing for drugs and alcohol in a person's or a race horse's blood or urine.
- Testing to see what is in illegal drugs that have been confiscated.
- Separating dyes and pigments in ink and paint.
- Testing for bombs at airports.
- Testing for the presence of ignitable fuels at a fire scene.

- Identifying fibers found at a crime scene.
- Separating newly discovered elements.
- Separating and testing histamines and antibiotics.
- RNA fingerprinting.
- Separating amino acids and anions.

What are pigments?
Pigments are molecules that absorb and reflect certain wavelengths of visible light. The color wavelengths they reflect are the colors we see. The other colors of the visible spectrum are absorbed. Pigments can be separated using chromatography.

Why doesn't permanent ink separate in water?
The pigments used in permanent pens are selected precisely because they do not dissolve in water. Because the pigments are not attracted to the water they simply are not moved as the water passes through them. These pigments can be dissolved, however, in some other solvents, such as alcohol.

What is in pen inks?
There are millions of different kinds of inks, and many more new ink formulas are invented every year! Companies may change their recipe from one year to the next if they find a less expensive or superior ingredient. One thing is true about all inks, however: They all have pigment(s) and a vehicle.

Pigment: A pigment is a substance that looks a certain color, because it reflects certain parts of the visible spectrum while absorbing others. It is a solid of some type that has been ground up into tiny pieces. It must be a solid that will not dissolve in the liquid vehicle, but will chemically bond to the substrate it is applied to. If it dissolves in the liquid, it is considered a dye, not a pigment. The molecules in dyes are smaller, which allows them to lie flatter on the substrate. This makes them reflect light more evenly, and appear more vivid, but also makes them more easily damaged by ultraviolet light. There are now some hybrid inks that include both pigments and dyes.

Vehicle: The vehicle needs to be a liquid that will not dissolve the pigments, but will flow well, carrying the pigments from the pen or printing press to the substrate. Once the vehicle reaches the substrate, it needs to be able to hold the pigment in place and thicken to a near-solid as quickly as possible. Although we often take ink for granted, it is difficult to find combinations of pigments and vehicles that work well together, and chemists are always trying to find better combinations.

Organic solvents are good at both flowing and hardening, and they were once used extensively in inks. However, because they contribute to air pollution, their use is now restricted. Resins, which originally came from trees, are used most often as ink vehicles, but they are now mostly synthetic. Oil is sometimes used.

Additives: These can be cleansing agents, moisturizing agents to prevent evaporation, buffering agents to control pH levels, fungicides, biocides, and resin for resilience.

What pigments can be seen when doing plant chromatography, and what purposes do they serve plants and people?

Pigments play important roles in plants. They are used for photosynthesis, metabolism, and to attract animals for pollination and seed dispersal. They are also important to humans and other plant-eating animals. They provide us with nutrients and attract our attention.

When we look at something, the color(s) we see represent the wavelengths of visible light that are reflected, not absorbed, by the object. Black absorbs all visible wavelengths of the spectrum. White reflects them all. Most plants look green, because chlorophyll reflects green, while absorbing red and blue. There are many other pigments in plants, but green chlorophyll is usually dominant, so we see them as green.

Depending on the solvents and mediums your students use, they may be able to see the pigments described below. Another way to see them is simply by looking at leaves during autumn. As the dominant green chlorophyll breaks down, you can see the yellow, orange, and red carotenes and xanthophylls. Chlorophyll and carotene are both fat-soluble and can be extracted from leaves and flowers with mineral oil. Mixtures of ether and acetone are also often used to separate plant pigments, but some procedures require careful safety measures.

Here are some of the pigments that can be seen when using chromatography on plants, and some information about them.

Chlorophyll a. The pigment's color is bright green, dark green, or blue green, so it reflects green wavelengths and absorbs red, blue, and violet. About 75% of the chlorophyll in plants is chlorophyll a. It plays the primary role in photosynthesis. Chlorophyll a is the only substance in nature that can capture light energy and convert it into chemical energy.

Chlorophyll b. The color is yellow green, light green, or olive green; it reflects these colors and absorbs red, blue, and violet. In photosynthesis, chlorophyll b plays an accessory role, absorbing light of a different wavelength than chlorophyll a, then transferring it to the chlorophyll a to be converted into chemical energy.

Carotenoids. This large group of chemicals includes two major groups of pigments: carotenes and xanthophylls. Both absorb the green light that chlorophyll does not. They are more stable than chlorophyll, and can be seen in leaves when the chlorophyll breaks down in autumn. **Carotenes** are found in all green plant tissue as well as in other plant parts, such as flowers and fruits. They reflect red, orange, and yellow wavelengths and absorb green and blue. They take in sunlight's energy and pass it to the chlorophyll. They also protect the plant from being harmed by ultraviolet light. Their nutrients include alpha-carotene, beta-carotene, and lycopene. Carotenes are found in yellow and orange fruits and vegetables, such as carrots, pumpkin, sweet potato, squash, apricots, mangoes, peaches, papaya, and cantaloupe. Lycopene is found in tomatoes and watermelon. Nutritional benefits to humans may include strengthening of the visual and immune systems, and preventing heart disease, breast, and cervical cancer. **Xanthophylls** reflect yellow wavelengths and absorb red, green, and blue. They too absorb sunlight energy and pass it to the chlorophyll and also offer protection from ultraviolet light. Nutrients include beta-cryptoxanthin, lutein, and zeaxanthin. They are found in dark green leafy vegetables, such as spinach, broccoli, collard greens, and kale, as well as in papaya, corn, red bell peppers, cilantro, oranges, and watermelon. They are thought to be helpful in the prevention of cataracts, heart disease, breast, and cervical cancer.

Anthocyanins. These reflect red and purple wavelengths and absorb blue, blue green, and green wavelengths. They also absorb ultraviolet rays and help prevent the plant from burning in strong light. If the sap of a plant is quite acidic, anthocyanins impart a bright red color; if the sap is less acidic, the color tends toward purple. In flowers and fruit, these pigments provide visual cues for animal pollinators and seed dispersers. Anthocyanins are found in strawberries, blueberries, purple grapes, plums, eggplant, red cabbage, red peppers, beets, and red apples. They are thought to be helpful in preventing cataracts and glaucoma and may delay cellular aging and prevent blood clots.

Brief Synopsis of Educational Research

The science education research literature includes a large body of work on alternate or mistaken conceptions that students of different ages may harbor on subjects related to the structure of matter. Chapter 15, The Research Base, in *Benchmarks for Science Literacy*, includes a summary of some of this research, including research on student ideas about the nature of matter, the conservation of matter, chemical changes, and the particle theory of matter. The last category is most relevant to this *Crime Lab Chemistry* unit and is quoted here:

> "Students of all ages show a wide range of beliefs about the nature and behavior of particles. They lack an appreciation of the very small size of particles; attribute macroscopic properties to particles; believe there must be something in the space between particles; have difficulty in appreciating the intrinsic motion of particles in solids, liquids and gases; and have problems in conceptualizing forces between particles (Children's Learning in Science, 1987). Despite these difficulties, there is some evidence that carefully designed instruction carried out over a long period of time may help middle-school students develop correct ideas about particles (Lee et al., 1993)."

Many of the strategies, approaches, and methods suggested by research as general guidelines for teachers in assisting their students in this specific area of scientific understanding are well represented in these revised *Crime Lab Chemistry* activities. Of course many of them also apply to other science learning. These include:

- Emphasize that science is about explanations; one of the ways that scientists make sense of the world is by constructing models or theories that are changed, improved, or refined as new evidence or understanding is gained.

- Use models to help students "visualize" the particulate nature of matter. Research suggests that teachers need to present the model, making sure students understand it and can visualize it, and show how the model can be used to explain certain phenomena. Practice using the model, challenge the model, and encourage students to see the limitations of the model. Teach a more sophisticated model as appropriate. In this unit, a number of different models are presented to students for their critique based on real phenomena, in order to focus them on the task of creating their own models.

- Be aware of alternate or naïve ideas or conceptions frequently held by students, and take notice if and when these ideas arise in class discussions or student work. Depending on the situation and timing, and with full respect for all student ideas, the teacher's challenge is to devise ways to allow the class to consider and reason out mistaken or limited aspects of these ideas over time, in order to move further along the continuum to improved understanding.

- Present a variety of phenomena to reinforce student learning. Of course, experiences with phenomena are not enough in and of themselves. Students need to engage in reflection, discussion, and argumentation in order to understand and appreciate how the phenomena relate to the scientific ideas. The teacher's task is to facilitate and guide, prompt, or focus that reflection.

- Elicit and strongly encourage students to express their reasoning, ask questions that probe for student thinking and prompt discussion. Once an idea begins to take hold, students (and all learners) still need many and varied opportunities to consider it from different sides and apply it.

- Provide students with practice in using new scientific ideas. Students need to have a sufficient number and variety of practice tasks to integrate ideas concerning particle theory. These should include tasks that ask students to develop descriptions and explanations of real-life phenomena.

- Ask students to demonstrate and apply the use of their newly gained knowledge through revision of their ongoing model, by analyzing a situation or image that seems to challenge the model, or other assessment and application assignments.

American Association for the Advancement of Science: Project 2061 (1993). *Benchmarks for Science Literacy,* Chapter 15, The Research Base, New York, Oxford University Press.

Brook, A., Briggs, H., and Driver, R. (1984). *Aspects of secondary students' understanding of the particulate nature of matter.* Children's Learning in Science Project, Leeds, UK: University of Leeds, Centre for Studies in Science and Mathematics Education.

Children's Learning in Science (1987). *Approaches to teaching the particulate theory of matter.* Leeds, UK: University of Leeds, Centre for Studies in Science and Mathematics Education.

Driver, R., Guesne, E., and Tiberghein, A., eds. (1985). *Children's Ideas in Science.* Philadelphia, Open University Press.

Lee, O., Eichinger, D.C., Anderson, C.W., Berkheimer, G.D., and Blakeslee, T.S. (1993). Changing middle school students' conceptions of matter and molecules. *Journal of Research in Science Teaching,* 30, 249–270.

Novick, S. and Nussbaum, J. (1978). Junior high school pupils' understanding of the particulate nature of matter: an interview study. *Science Education* 62(3): 273–281.

Novick, S. and Nussbaum, J. (1981). Pupils' understanding of the particulate nature of matter: a cross-age study. *Science Education* 65(2): 197–196.

Scott, P. (1987). The process of conceptual change in science: a case study of the development of a secondary pupil's ideas relating to matter. In *Proceedings of the Second International Seminar on Misconceptions and Educational Strategies in Science and Mathematics* (Vol. II: 404–419), ed. J.D. Novak. Ithaca, NY, Department of Education, Cornell University. ■

TEACHER'S OUTLINE

ACTIVITY 1: LABORATORY INVESTIGATIONS

Session 1: Investigating the Evidence

■ Getting Ready

1. Choose, adapt, or make up mystery scenario.
2. Copy **Teacher Script** and **Suspect Statements.** Cut apart statements.
3. Prepare materials as described in guide.

■ Introducing the Mystery

1. From the script, read aloud "Introducing the Mystery" including the ransom note.
2. Say note written on white paper towel with black felt-tip pen. Six suspects have been identified. Point out pens but don't introduce suspects yet.
3. Ask for ideas about how to find out which pen was used. Accept all ideas.
4. Write ***chromatography*** on board. Say it's a technique chemists use in crime labs, and that students will use it to help solve the mystery.

■ Conducting a Blank Test

1. Use a *blank* paper strip to demonstrate the technique.
2. Have students make predictions, and encourage discussion.
3. Distribute materials and have them begin.
4. Circulate; encourage close observation.

■ Testing the Ransom Note

1. Read note aloud again.
2. Hold up paper towel strip of note. Ask for and record predictions about what will happen as water travels up through ink.
3. Remind students this fragment of note is their *only* piece of evidence. Caution them not to let ink line dip below water's surface, to remove strip when water has traveled about three-quarters up, then tape it to paper to dry.
4. Distribute a piece of note to each pair and have them conduct test.
5. Reconvene class. Have students share observations.
6. Explain that the colors they see are pigments mixed to make black ink; a ***pigment*** is a chemical that makes something look a certain color; and ink is a mixture of different substances.
7. Further explain that chromatography is a technique that separates ***substances*** from a ***mixture.*** Each different mixture produces its own color pattern (called a ***chromatogram***).
8. Ask a student to describe color at bottom of chromatogram. Ask if anyone disagrees. Continue with each color in chromatogram.
9. If disagreement about colors, ask why groups might have different results. Introduce ***variable*** as something that can change from one test to next.

■ Introducing the Suspects

1. Post or reveal list of suspects. Choose volunteers to read the parts of suspects. Pass out suspect statement to each student actor.
2. Read "teacher" lines from "Introducing the Suspects" section of script and have students read their assigned parts.
3. As each suspect is introduced, point out name and pen number on list.

■ Testing the Mystery Pens

1. Ask students how they could use chromatography to discover which suspect's pen was used to write the note.
2. Describe procedure as in guide. Have students begin.
3. As students finish, have them tape pen chromatograms to paper next to note strip and write their names on the paper.
4. When everyone finishes, have volunteers collect materials, including papers of chromatograms which are laid out to dry. Say that in next class session, they'll discuss test results.

Session 2: Solving the Mystery

■ Getting Ready

1. Make overhead of **Different Water Molecule Models.**
2. Have on hand students' chromatograms and list of suspects.
3. Have drawing paper and pencils ready. If using lenses, set them out.

■ Identifying the Mystery Pen

1. Pass out chromatograms (and lenses). Ask students to examine chromatograms to determine which pen was used to write note.
2. Encourage students to examine another group's chromatograms to see if evidence is consistent.
3. Reconvene the class. Ask if students know which is the "mystery pen." Have them present evidence.
4. Continue any discussion until agreement about pen is reached. Refer to list of suspects as needed.

■ Does the Evidence Prove the Suspect is Guilty?

1. Ask if they think they have enough evidence to convict suspect.
2. Ask who thinks the matching pen is *not* enough evidence to convict.
3. Acknowledge that one explanation for the evidence is that suspect is guilty. Challenge students to explain how suspect could be innocent. Have student pairs discuss.
4. Reconvene the class and ask students to explain their ideas.
5. After a thorough discussion, ask students to vote again on whether or not the matching pen is enough evidence to convict. Say that in the real world, more evidence would be needed.

■ The "Microscope Eyes" Model of Chromatography

1. Ask for observations about how the pen chromatograms compare.
2. Ask why some pigments traveled higher than others.
3. Say that all substances (including water, ink pigments, paper) are made up of atoms and ***molecules***—extremely tiny particles. Different substances have molecules of different shapes, sizes, and weights.
4. Draw classic "Mickey Mouse" version of a water molecule on board. Say a water molecule consists of two hydrogen atoms bonded with one oxygen atom.
5. Although we can't see water molecules, scientists draw them this way based on evidence of how water behaves. It's a model.
6. Show the transparency. Say that each of these models has strengths and weaknesses, showing some things well, and other things not as well. To be *completely* accurate models would have to be the real thing.
7. Ask students to pretend they have "microscope eyes" and can see what ink, paper, and water molecules look like.
8. Say they'll draw a picture or "model" of how they imagine the molecules might be shaped. Their drawing should try to show how the shapes, interactions, or other characteristics of the water, ink, and paper molecules might explain why pigments are carried up the paper and why some are carried higher than others. Say drawing is just a first draft and needn't be "correct."
9. Pass out paper and pencils. Ask students to write name on paper and label it as "Microscope Eyes Model of Chromatography." Have them begin drawing.
10. Circulate and encourage students to note what parts of chromatography or their model they are unsure of, or what their model explains well.
11. Ask some students to share their drawings. Reiterate that these are models—they *represent* something. Either collect drawings or have students keep them to make adjustments in the next activity.

ACTIVITY 2: WHAT IS CHROMATOGRAPHY?

■ Getting Ready

1. Choose a long section of *non-carpeted* floor for demonstration of model chromatography system. Plug in fan or blow dryer and prepare/gather material for demonstration. Test the model before class.
2. Make overheads of **Model A, Model B, Model C, Model D,** and **Four Models,** and one copy for each student of **Four Models** and **New Microscope Eyes Model of Chromatography.**
3. Have pencil and "Microscope Eyes Model of Chromatography" drawings ready to distribute. If necessary, plan ahead how to re-group students for "Revisiting the 'Microscope Eyes' Model" part of activity.

■ Explaining the Chromatography System

1. Sketch chromatography set up used in Activity 1.
2. Remind students chromatography is used by many different kinds of chemists to separate mixtures into their parts. To understand how it works, it's helpful to analyze how chromatography is a system.

3. Define a ***system*** as a group of interacting parts that work together. Say in the chromatography system there are always three main parts—the ***test substance, medium,*** and ***solvent.*** Label each part of the system as you ask questions about and define them.
4. Ask why ink from one of the pens didn't travel up the paper at all. Ask if another solvent would make the permanent ink molecules travel up the paper.
5. Explain that there are many types of solvents and mediums used in chromatography, and that it is an incredibly important process and tool in chemistry, because it can be used to separate the parts of almost any substance.

■ Demonstrating a Model Chromatography System

1. Gather students around floor area you have chosen and say you'll demonstrate how chromatography works by using a large model.
2. Ask the students to imagine that they're chemists who want to separate the parts of a test substance.
3. Set the materials in front of the fan and conduct the demonstration, as described in guide, being sure to use the terms test substance, medium, and solvent. Point out how each is represented in the model.
4. Have students return to their seats.

■ Revisiting the "Microscope Eyes" Model

1. Refer back to your drawing to briefly summarize how paper chromatography system works.
2. Say that since students now know more about chromatography, they'll analyze four drawings. Each is a possible model for how the ink molecules separated in the chromatography test. **Emphasize that all four models have some things that are accurate and some things that are not accurate.**
3. Show transparencies of chromatography **Models A** through **D** and *briefly* explain each as detailed in guide.
4. Show transparency of **Four Models** student sheet and say they'll each get a sheet like this.
5. Explain that after many years of scientific investigation, scientists have learned some facts about molecules. Go over "Things we know about molecules." Encourage students to re-read these facts as they think about each model.
6. Pass out **Four Models** sheets. Say students should read them, discuss in their group what is inaccurate or accurate about each model, then write answers on their own sheet.
7. After students have discussed and recorded their answers, hold a brief discussion of what is accurate or inaccurate about each model, as described in guide.

■ Revising Their Earlier Models

1. Review the idea that separation of molecules has to do with their size, weight, and shape and how strongly they are attracted to solvent or medium. Scientists don't have the perfect model of what goes on at the molecular level; there's always room for improvement.
2. One sign of a good scientist is the ability to change ideas when presented with new evidence.
3. Say you'll hand back their "Microscope Eyes Model of Chromatography" drawings from Activity 1. Their challenge is to make a new drawing incorporating what they now know about the chromatography system and molecules.
4. Hold up **New Microscope Eyes Model of Chromatography** sheet. Say students will draw what

they think the molecules might look like in the appropriate places on the sheet.

5. Say their model should show their current idea of what the different ink pigments look like and explain why some go higher on the paper than others. They can refer to their earlier drawing and the **Four Models** sheets.
6. Give students their first drawing, then pass out **New Microscope Eyes Model of Chromatography** sheet. Have students make their new drawings.

■ (Optional) Sharing the New Microscope Eyes Models

1. Assign students to groups of four. Give each student a pencil and a sheet of paper to be labeled "Questions." Have them write their name on the Questions sheet, and on sheet with new drawing.
2. Tell each student to pass these two papers to the person sitting to their left in their group. Ask them to study the drawing handed to them, try to figure it out, and write any questions they have about it on the accompanying Questions sheet.
3. After a few minutes, tell them to pass the papers to the left again and follow the same procedure. Continue until every student gets back his or her own papers.
4. Give students a few minutes to read and think about the questions written by other students. Tell them to think of how they might want to readjust their drawing after reading the questions.
5. Tell the groups to take turns discussing the drawings of each of their members one at a time. They can describe each model, ask questions (those they wrote or new ones), and respond to the questions.

ACTIVITY 3: DESIGNING YOUR OWN TESTS

Session 1: Choosing a New Test Substance

■ Getting Ready

1. Choose a long counter or long table for the Test Substances station. Spread newspaper or plastic bags to protect the surface.
2. Gather a variety of substances for the station, as described in guide, and set the materials at the station.
3. Make one copy each of the **Candy Coating, Food Coloring,** and **Plants Test Substance Procedure** sheets. Place each near the appropriate substances at the station.
4. Decide how desks will be arranged so students can share materials in groups of four to six.
5. Prepare material for each group, as described in guide. Have it ready for distribution.

■ Introduce the Activity

1. Briefly review what chromatography is used for and the parts of a chromatography system.
2. Explain they'll again use water as a solvent and strips of chromatography paper as the medium. Ask what substances they could test other than inks from pens.
3. Show students items at Test Substances station. **Emphasize they don't need to test all substances** but should test a few that interest them, **with the goal of finding one test substance to investigate further** in next session.

4. Say everyone in group will share water trough, strips of chromatography paper, pencils, tape, and scratch paper, but chromatography tests will be done with a partner. Assign partners.
5. As in guide, explain procedure and show how to apply different test substances. Mention procedure signs.
6. Describe the calm, cooperative behavior you expect of students.

■ Students Test Substances

1. Distribute materials to groups and allow students to begin. Circulate to make sure students follow procedure.
2. End when all groups have tested at least three substances. Collect materials from groups and collect papers with chromatograms taped on them. Have someone straighten up the Test Substances station, and leave it set up for the next session.
3. Tell class they'll share their results and do further tests in next session.

Session 2: Designing Chromatography Systems

■ Getting Ready

1. Make one copy of **Our Chromatography Test** for each *pair* of students, and one copy of **Chromatography Test Report** for *each student.*
2. Have handy the tape, pencils, and papers with chromatograms.
3. Prepare materials and set up the Test Substances, Mediums, and Solvents stations as described in guide.

■ Discuss Test Substance Results

1. Distribute papers with chromatograms from previous session. Ask a few students to share their results.
2. Ask what substances didn't separate at all or were difficult to see, and what could be done differently to get better separation.

■ Introduce the Activity

1. Explain that, based on their results, each pair will choose **one test substance.** They will then design their own chromatography system with the best medium and solvent to get the best separation they can.
2. Ask for ideas on what mediums they might use other than chromatography paper. Show them test strips available at Mediums station.
3. Ask what solvents they might use other than water. Point out what's available at the Solvents station.
4. If you've added permanent pens to the Test Substances station, let students know about them.

■ Designing Chromatography Systems

1. Tell students they'll start by looking over their chromatograms, then decide with their partner on **one** test substance to test further.
2. Hold up an **Our Chromatography Test** sheet. As described in guide, briefly demonstrate procedure using a student as your partner.

3. Hand out sheets to each pair, and give tape and a few pencils to each group.
4. Once pairs have chosen and recorded their test substance, allow them to go to stations.
5. Circulate to ensure partners are working together, following the test procedures, and recording their results after each test.

■ Discussing and Reporting Results

1. When students have finished their tests and completed questions on sheets, reconvene for a class discussion.
2. Have a few students share what test substance they chose, what mediums and solvents they tried, and why they think they got the results they did. What were some surprises? Frustrations? Did they need to change their focus midway?
3. Say that each student will now get a **Chromatography Test Report** on which to record the details of what she found out. Go over the questions on the sheet, and emphasize that explanations need not be "correct," but do need to be based on evidence.
4. Point out the place where they'll draw a "microscope eyes" model that shows what they think happened to the molecules of the test substance, medium, and solvent during their most successful test.
5. Pass out a sheet to each student and have them begin.
6. Have early finishers clean up and organize the station materials.
7. When students are finished, collect the **Chromatography Test Report** sheets.
8. As appropriate, after you've reviewed their work, discuss their responses, clarify misunderstandings, and encourage students to raise ongoing questions.

Assessment Suggestions

Anticipated Student Outcomes

1. Students are able to explain that chromatography is a technique used to separate mixtures into their separate ingredients. They understand that the "colors" they see in a chromatogram are the different substances/pigments that were mixed together to make the test substance.

2. Students are able to articulate that chromatography can be understood as an interacting system, involving a test substance, a solvent, and a medium.

3. Students demonstrate their ability to use paper chromatography to separate substances from a mixture and to devise their own combinations of test substance/solvent/medium.

4. Students gain increased insight into the particulate nature of matter and are able to create and refine molecular models to represent their own thinking about why a test substance, solvent, and medium behave as they do.

Embedded Assessment Activities

Probing Questions. At various points throughout the activities, the teacher asks key questions such as: "Where do the colors come from?" "Which ink contained the greatest number of pigments?" "Why might different groups have gotten different results?" "Why didn't pen # 6 move up the paper?" Student responses can reveal their understanding. Some teachers like to ask questions such as these more than once during the unit, thereby seeing students' progress in understanding. (Addresses outcomes 1, 2, 3, 4)

Microscope Eyes Model. At the end of Activity 1 and in Activity 2, students are introduced to the idea of molecular models and begin creating their own. They also critique four possible models that are presented to them. This is a very rich assessment platform. Teachers can gain insight into students' initial understandings when they look at the early models students make. They can assess student growth as they look at the ways students change, refine, and discuss/debate their models over the course of the unit. Encouraging discussion and debate and having students express their underlying reasoning can reveal the accuracy or inaccuracy of their underlying conceptions and the aspects where further instruction may be needed. (Outcomes 2, 4)

Our Chromatography Test and the **Chromatography Report.** These two student sheets provide important assessment information in a number of ways. Among the abilities and understandings a teacher can assess are: student ability to explain chromatography as an interacting system, to originate their own combinations of test substance, solvent, and medium, and to further refine their molecular models. These sheets also can reveal student understandings about inquiry, scientific test design, and the scientific method. (Outcomes 2, 3, 4)

Colored Pens and Plant Pigments. During the last activity, students use chromatography to analyze the pigments found in different substances. In one activity, students determine what pigments were combined to create the colors of ink in a variety of watercolor markers. In another, they observe which solvents cause plant pigments to move up the test strip, and how many different pigments can be separated from the plant. The teacher can observe the students' ability to perform chromatography tests and interpret the resulting chromatograms. The plant pigment activities offer the teacher an opportunity to observe how well students can apply their understanding of solubility and solvents. (Outcomes 1, 2, 3, 4)

Additional Assessment Ideas

Going Furthers. A number of the "Going Further for the Whole Unit" suggestions on pages 63–64 result in student work that teachers could use to assess understanding. If you do use, for example, a student-originated mystery plot as an assessment, make sure to tell students the main criteria you will use in assessing their work. (Outcomes 1, 2)

Tell the Story of a Chromatogram. Show students a chromatogram made from black ink. Ask them to explain how it was made, where the colors come from, what the colors are, and what the colors tell us. Show a second chromatogram made from the same substance that looks similar but has differences. Explain that this is a chromatogram of the same ink, made by a different person. Ask students to explain how two chromatograms of the same ink could have different appearances. (Outcomes 1, 3)

RESOURCES AND LITERATURE CONNECTIONS

Sources

Classroom-Ready Materials Kits

Carolina Biological Supply® is the exclusive distributor of fully prepared GEMS Kits®, which contain all the materials you need for full classroom presentation of GEMS units. For more information, please visit www.carolina.com/GEMS or call (800) 227-1150.

Chromatography Paper

Chromatography paper can be purchased from many scientific supply stores, including Carolina Biological Supply.

Carolina Biological Supply
2700 York Rd.
Burlington, NC 27215-3398
(800) 334-5551
www.carolina.com

Whatman No. 1 Chromatography Paper
Available in 1 1/2 inch x 300 foot rolls and 12 cm square sheets.

Flinn Scientific, Inc.
P.O. Box 219
Batavia, IL 60510-0219
(800) 452-1261
www.flinnsci.com

Available in 1 inch x 600 foot rolls and 8 inch square sheets.

Sargent-Welch
P.O. Box 5229
Buffalo Grove, IL 60089
(800) 727-4368
www.sargentwelch.com

Whatman No. 1 Chromatography Paper
Available in 18 inch x 22 1/2 inch sheets and rolls of different lengths and widths.

Whatman No. 4 Chromatography Paper
Slightly heavier than Whatman No. 1. Available in 46 cm x 57 cm sheets.

Ward's
5100 West Henrietta Rd.
P.O. Box 92912
Rochester, NY 14692-9012
(800) 962-2660
www.wardsci.com

Available in 3/4 inch x 6 inch strips.

Related Curriculum Material

Science and Technology Concepts for Middle Schools™ (STC/MS™)

An inquiry-based middle school (grades 6–8) science curriculum developed by the National Science Resources Center. It contains eight modules in four science strands—Life Science, Earth Science, Physical Science, and Technology. In the Properties of Matter module, students investigate characteristic properties of matter such as density, boiling and freezing points, and solubility, then utilize these properties to identify substances. Later they use the characteristic properties to separate mixtures into pure substances, and investigate the interaction of compounds and elements.

Nonfiction for Students

Women in Chemistry
Their Changing Roles from Alchemical Times to the Mid-Twentieth Century
by Marelene Rayner-Canham and Geoffrey Rayner-Canham
Chemical Heritage Foundation, Philadelphia, PA
(2001; 298 pp.)

This highly readable book sheds new light on the past by tracing the life, times, and scientific contributions of more than 50 women chemists. Compelling biographies of medieval alchemists and Nobel Laureates alike are presented in scientific and cultural contexts. Social barriers to formal scientific education of women are identified throughout the centuries, as well as the importance of mentors.

Fiction for Students

Note: Naturally, most of the books are mysteries, and as such concern forensic testing of evidence and the important distinction between evidence and inference, thus making a strong connection to the *Crime Lab Chemistry* unit. We list only a few of our favorites; we're sure you and your students have many others.

From the Mixed-Up Files of Mrs. Basil E. Frankweiler
by E.L. Konigsburg
Atheneum, New York, NY
(1967; 162 pp.)
Grades 5–8

Twelve-year-old Claudia and her younger brother run away from home to live in the Metropolitan Museum of Art and stumble upon a mystery involving a statue attributed to Michelangelo. This book is a classic, and has been recommended to GEMS by many teachers. The detecting techniques referred to include chromatography.

The Great Adventures of Sherlock Holmes
by Arthur Conan Doyle
Viking Penguin, New York, NY
(1990; 264 pp.)
Grades 6–Adult

These classic stories are masterly examples of deduction. Many cases are solved by Holmes in his chemistry lab as he analyzes inks, tobaccos, mud, and other substances. Available from many publishers in many editions.

The Missing Gator of Gumbo Limbo
An Ecological Mystery
by Jean C. George
HarperCollins, New York, NY
(1992; 160 pp.)
Grades 4–7

Sixth-grader Liza K. and her mother live in a tent in the Florida Everglades. She becomes a nature detective while searching for Dajun, a giant alligator who plays a part in a waterhole's oxygen-algae cycle, but is marked for extinction by local officials. Liza K. is motivated to study the ecological balance by her desire to keep her outdoor environment beautiful. This "ecological mystery" combines precise scientific information and environmental concerns with good characterization, a strong female role model, and a compelling plot. Another recommended book by the same author is *Who Really Killed Cock Robin?* (HarperCollins, 1991).

Motel of the Mysteries
by David Macaulay
Houghton Mifflin Co., Boston, MA
(1979; 96 pp.)
Grades 6–Adult

Presupposing that all knowledge of our present culture has been lost, an amateur archaeologist of the future discovers clues to the lost civilization of "Usa" from a supposed tomb, Room #26 at the Motel of the Mysteries, which is protected by a sacred seal (a "Do Not Disturb" sign). This book is an elaborate and logically constructed train of inferences based on partial evidence, in a pseudo-archaeological context.

Reading this book, whose conclusions they know to be askew, can encourage students to maintain a healthy and irreverent skepticism about their own and other's inferences.

The Mysteries of Harris Burdick

by Chris Van Allsburg
Houghton Mifflin, New York, NY
(1984; 32 pp.)
Grades 2–5

Presents a series of loosely related drawings—each accompanied by a title and a caption—which the reader may use to make up his or her own story.

The Real Thief

by William Steig
Farrar, Straus & Giroux, New York, NY
(1973; 64 pp.)
Grades 4–8

King Basil and Gawain, devoted Chief Guard, are the only two in the kingdom who have keys to the Royal Treasury. When rubies, gold ducats, and finally the world-famous Kalikak diamond disappear, Gawain is brought to trial for the thefts. But is he the real thief? As the mystery unfolds, it becomes clear that it is important to investigate fully before making judgments or drawing conclusions.

Sammy Keyes and the Hotel Thief

by Wendelin Van Draanen
Knopf, New York, NY
(1998; 168 pp.)
Grades 4–7

Thirteen-year-old Sammy's penchant for speaking her mind gets her in trouble when she involves herself in the investigation of a robbery at the "seedy" hotel across the street from the seniors' building where she is living with her grandmother. The first in a series of books featuring Sammy, a saucy girl detective.

Time Stops for No Mouse
A Hermux Tantamoq Adventure

by Michael Hoeye
G. P. Putnam's Sons, New York, NY
(2002; 250 pp.)
Grades 5–8

When Linka Perflinger, a jaunty mouse, brings a watch into his shop to be repaired and then disappears, Hermux Tantamoq is caught up in a world of dangerous search for eternal youth as he tries to find out what happened to her. A well-told tale in which suspense and fantasy are intertwined.

The Westing Game

by Ellen Raskin
Avon, New York, NY
(1984; 192 pp.)
Grades 6–10

The mysterious death of an eccentric millionaire brings together an unlikely assortment of 16 beneficiaries. According to instructions contained in his will, they are divided into eight pairs and given a set of clues to solve his murder and thus claim the inheritance. Newbery award winner. Another recommended book by the same author is *The Tattooed Potato and Other Clues* (E.P. Dutton, 1975; Penguin, 1989).

Video

Private Universe Project in Science

A video workshop on teaching science for grade K–12 educators. It explores why teaching science is so difficult and offers practical advice to teach more effectively. In Workshop 4 "Chemistry: A House With No Foundation," the video shows elementary, middle, and high school teachers comparing students' ideas about the particulate nature of matter. Available through Annenberg/CPB. For more information go to www.learner.org/resources/series29.html

Internet Sites

Note: While we do our best to provide long-lived addresses in this section, websites can be mercurial! Comparable alternative sites can generally be found with your Web browser.

Black Magic
www.exploratorium.edu/science_explorer/black_magic.html

An "at home" chromatography activity, including an explanation of why inks separate.

Strange Matter
www.strangematterexhibit.com/index.html

Developed for grades 5–8, this site features a collection of virtual activities for students to explore materials science. From the homepage, students can choose among four activities to discover the science behind everyday materials. The site also contains information for teachers, including a teacher's guide, a featured demonstration, curriculum connections, and other resources.

Reviewers

We warmly thank the following educators, who reviewed, tested, or coordinated the trial tests in manuscript or draft form. Their critical comments and recommendations, based on classroom presentation of these activities nationwide, contributed significantly to this GEMS publication. (The participation of these educators in the review process does not necessarily imply endorsement of the GEMS program or responsibility for statements or views expressed.) Classroom testing is a recognized and invaluable hallmark of GEMS curriculum development; feedback is carefully recorded and integrated as appropriate into the publications. WE THANK THEM ALL! ■

ALASKA

Galena City School, Galena
Olyn Garfield*

ARIZONA

Acacia School, Phoenix
Mark Kauppila

Alta Vista School, Phoenix
Jean Reinoehl

Cactus Wren School, Phoenix
Tim Maki

Cholla School, Phoenix
Brenda Pierce
Shirley Vojtko

Desert Foothills School, Phoenix
Liz Sandberg

John Jacobs School, Phoenix
George Lewis
John O'Daniel

Lookout Mountain School, Phoenix
Flo-Ann Barwick
Charri Strong

Manzanita School, Phoenix
Edie Helledy
Sandy Stanley

Maryland School, Phoenix
Greg Jesberger

Moon Mountain School, Phoenix
Bill Armistead
Karen Lee
Don Metzler

Orangewood School, Phoenix
Donna Pickering

Palo Verde School, Phoenix
John Little
Tom Lutz

Roadrunner School, Phoenix
Bob Heath

Washington School, Phoenix
Ken Redfield

Washington School District, Phoenix
Richard E. Clark*

CALIFORNIA

Albany Middle School, Albany
Susan Butsch
Robin Davis
Joanna Klaseen
Cindy Plambeck
Susan Power
Rich Salisbury
Bob Shogren*
Kay Sorg

Cornell School, Albany
Susan Chan
Pamela Zimmerman

Vista School, Albany
Bob Alpert*

White Hill Junior High School, Fairfax
Karen Ardito

Horner Junior High School, Fremont
Claudia Hall
Linda McClanahan*

Frick Junior High School, Oakland
Mark Piccillo
Theodore L. Smith

Sleepy Hollow School, Orinda
Margaret Lacrampe

Cave Elementary School, Vallejo
Dale Kerstad★
Tina Neivelt
Neil Nelson
Carol Rutherford
Jim Salak
Bonnie Square

Dan Mini Elementary School, Vallejo
James Boulier
Jack Thornton★

Vallejo City Unified School District, Vallejo
Secondo Sarpieri★
Alice Tolinder★

KENTUCKY

Adath Jeshurun Preschool, Louisville
Mary Artner
Sandi Babbitz
Riva Drutz
Linda Erman
Jennie Ewalt
Ann Peterson
Harriet Waldman

Brown School, Louisville
Tony Peake
Larry Todd

DuPont Manual Magnet School, Louisville
Joan Stewart

Holy Spirit School, Louisville
Phyl Breuer

Jefferson County Public Schools, Louisville
Brad Matthews
Ken Rosenbaum

Museum of History and Science, Louisville
August Drufke
Sam Foster
Amy S. Lowen★
Theresa H. Mattei★
Mike Plamp
Edna Schoenbaechler
Melissa Shore
Dr. William M. Sudduth★
Fife Scobie Wicks
August Zoeller
Doris Zoeller

Prelude Preschool, Louisville
Sherrie Morgan

Sacred Heart Model School, Louisville
Laura Hansen
Pam Laveck
Sister Mary Mueller

St. Francis High School, Louisville
Susan Reigler

Thomas Jefferson Middle School, Louisville
Toni Davidson
Nancy Glaser
Leo Harrison
Muriel Johnson
Cathy Maddox
John Record
Jenna Stinson

Wheatley Elementary School, Louisville
Alice Atchley
Anne Renner

MICHIGAN

Harper Creek High School, Battle Creek
David McDill
Rebecca Penney

Gagie School, Kalamazoo
Barbara Hannaford
Susie Merrill
Sue Schell
Julie Schmidt

Kalamazoo Public Schools, Kalamazoo
Gloria Lett★

Lincoln Elementary School, Kalamazoo
Tina Echols

Northeastern Elementary School, Kalamazoo
Kathy Patton

Science and Mathematics Education Center Western Michigan University, Kalamazoo
Dr. Alonzo Hannaford★
Rita Hayden★
Dr. Phillip T. Larsen★
Rick Omilian★

South Christian School, Kalamazoo
Dave Bierenga
Edgar Bosch

South Junior High School, Kalamazoo
Deb Ply

Portage Central High School, Portage
Ruth James

Woodland Elementary School, Portage
Joann Dehring
Mary Beth Hunter
Roslyn Ludwig
Everett McKee
Joel Schuitema
Bev Wrubel

Schoolcraft Elementary School, Schoolcraft
Shirley Pickens
Sharon Schillaci

Schoolcraft Middle School, Schoolcraft
Craig Brueck

NEW YORK

Discovery Center, Albany
Sigrin Newell

Albert Leonard Junior High School, New Rochelle
David Selleck

City School District of New Rochelle, New Rochelle
Dr. John V. Pozzi*

Columbus Elementary School, New Rochelle
Rubye Vester

George M. Davis Elementary School, New Rochelle
Edna Neita
Julia Taibi

Isaac E. Young Junior High School, New Rochelle
Bruce Zeller

Jefferson Elementary School, New Rochelle
Tom Mullen

Trinity Elementary School, New Rochelle
Frances Bargamian
Bob Broderick

Ward Elementary School, New Rochelle
Eileen Paolicelli
John Russo
Tina Sudak

Webster Magnet Elementary School, New Rochelle
Richard Golden*
Bruce Seiden

NORTH CAROLINA

Discovery Place, Charlotte
Ed Gray
Sue Griswold
Mike Jordan
John Paschal
Cathy Preiss
Carol Sawyer
Patricia J. Wainland*

North Carolina Museum of Life and Science, Durham
Jorge Escobar
James D. Keighton*
Paul Nicholson

OHIO

Miami University, Middletown
A.M. Sarquis

OREGON

Oregon Museum of Science and Industry
Christine Bellavita
Judy Cox
David Heil*
Shab Levy
Joanne McKinley
Catherine Mindolovich
Margaret Noone*
Jim Todd
Ann Towsley

Oregon Museum of Science and Industry (OMSI) staff conducted trial tests at the following sites:
Berean Child Care Center, Portland
Grace Collins Memorial Center, Portland
Mary Rieke Talented and Gifted Center, Portland Public School District, Portland
Portland Community Center, Portland
Portland Community College, Portland
St. Vincent De Paul, Child Development Center, Portland
Salem Community School, Salem
Volunteers of America, Child Care Center, Portland

WASHINGTON

Pacific Science Center, Seattle
David Foss
Stuart Kendall
Dennis Schatz*
William C. Schmitt
David Taylor

FINLAND

Katajanokan Ala-Aste, Helsinki
Pirjo Tolvanen

Katajanokka Elementary School, Helsinki
Arja Raade
Gloria Weng*

Åbo Akademi, Vasa
Sture Björk

*Trial test coordinators